New Perspectives in Caribbean Tourism

Routledge Advances in Tourism

New Perspectives in Caribbean Tourism

Edited By

Marcella Daye,
Donna Chambers,
and
Sherma Roberts

 Routledge
Taylor & Francis Group

New York   London

First published 2008
by Routledge
711 Third Avenue, New York, NY 10017

Simultaneously published in the UK
by Routledge
2 Park Square, Milton Park, Abingdon, Oxfordshire OX14 4RN

Routledge is an imprint of the Taylor & Francis Group, an informa business

First issued in paperback 2011

© 2008 Taylor & Francis

Typeset in Sabon by IBT Global.

Library of Congress Cataloging-in-Publication Data
New perspectives in Caribbean tourism / edited by Marcella Daye, Donna Chambers, and Sherma Roberts.
    p. cm. — (Routledge advances in tourism)
    ISBN 978-0-415-95838-7
    1. Tourism—Caribbean Area. I. Daye, Marcella. II. Chambers, Donna. III. Roberts, Sherma.
    G155.C35N49 2008
    338.4'791729—dc22   2007042962

ISBN13: 978-0-415-95838-7 (hbk)
ISBN13: 978-0-415-89786-0 (pbk)
ISBN13: 978-0-203-93127-1 (ebk)

Contents

PART II
Governance

List of Tables and Figures

TABLES

FIGURES

Foreword

New Perspectives in Caribbean Tourism could not be more aptly named as a text for our times and for the foreseeable future. The grudging acceptance of tourism as a tool for economic development in the Caribbean and the emerging view of tourism as simply that part of GDP that is derived from the economic activities of visitors are coming amidst the elevation of all discussions from one of "boon or bane" to a full recognition that tourism is most probably both.

We are also coming to understand that tourism is not a career; it is not an industry. Instead, it is a sector that pervades our entire economy, and, because it has such a considerable economic and sociological impact on our Caribbean states, it is deserving of much deeper study than it has been accorded to date. There is no question that a full examination of tourism has been retarded by the early declaration from several Caribbean luminaries about the "evils" of tourism. Many held on to the view of tourism as little more than the sons and daughters of former slaves serving the sons and daughters of former slaveholders, which delayed any embrace of tourism as an economic-development tool even as it was clear that the Caribbean held a natural global competitive and comparative advantage in that area. That view also delayed subjecting tourism to the kind of intellectual rigour that industries such as agriculture received through the years and, except for funds for promotion, delayed the embrace of tourism by many governments throughout our region. This view accounts probably for the general avoidance of the core jobs in the sector by many of our region's best, brightest, and most energetic citizens. This last result is most unfortunate because the largest sector of Caribbean economies has developed without competition to get into the sector from the region's most talented citizens.

There are those who are concerned that any review of the sociological impact of tourism on our societies will retard economic development, but there are also many who believe that some elements of tourism are clearly having a favourable impact on our societies. For example, there is little doubt that much of the concern for the environment is moving interest from developing ecotourism to developing "eco-nations" far faster than would have been the case without the growing concern from the tourism sector about the environment. There are also some who point to the possibility of delivering a higher quality of more locally and regionally grown products to the tables of our visitors and residents and thereby revive the flagging agricultural sector. The new e-commerce tools that are now available also promise maintenance of easy access to and easy delivery of some Caribbean products and services to past visitors at their normal place of residence long after their trip to the region, thus extending the business and cultural relationship over a longer period of time. So tourism is indeed a two-edged sword. None of this escapes the view of the authors.

It is important that these "new perspectives" are being introduced by a relatively youthful group of writers, contributors, and editors. Their thoroughness in covering the subject is quite evident, and the tenor of our times might permit this book to become the catalyst for ongoing discussion of tourism. Now that travel and tourism are generally accepted as the largest sector of the global economy, there is growing interest in the Caribbean about the proper management of the sector in many progressive states even while they fight for the sustainability of local customs and cultures. This growing interest is accelerated by the recognition that tourism is that one sector that affects and is affected by everyone in the society. So an unintended but welcome consequence of this full ranging review is a "new perspective" on the evolving Caribbean.

Vincent Vanderpool-Wallace
Secretary General of the
Caribbean Tourism Organisation,
St. Michaels, Barbados

Preface

Growing up in postindependent Caribbean countries, the editors of this volume have been educated to view the region's natural, cultural, and social characteristics as celebratory. So that in spite of the traumas and the systemic inequality and subjugation that were etched into the landscape of plantation economy and society, we were schooled to orient our thinking to valorise the emancipatory processes of resistance demonstrated not only in moments of rebellion but also in the survival strategies of syncretisation and adaptation as a means of achieving development and self-determination. Yet this perspective of our Caribbean identity is not an attempt to romanticise or construct the region as a "master code" or an ideal type because we are fully aware of the realities of the region's internal tensions and conflicts, as well as the peripherality and dependency of our societies today in global geopolitics. Nevertheless, the grammar of identity that inscribes our thinking, formalised in our undergraduate and postgraduate education at the University of the West Indies, has undeniably been expressed in our academic proclivities and interests.

However, our entry into the world of Caribbean tourism came first in practice having all been employed in the industry at the level of marketing, policy, and hospitality. Our quest for further training, individually, led us to the University of Surrey in the United Kingdom, where we embarked on postgraduate studies in tourism. It was in the pursuit of both master's and doctoral studies in the United Kingdom that we encountered what we considered to be a dearth of research on Caribbean tourism, particularly research conducted by Caribbean nationals or within the tradition of Caribbean scholarship. It was in recognition of this gap in tourism scholarship that this volume was conceived. We consider that this effort is still embryonic, but it should be nurtured and perhaps evolve into a distinct body of tourism knowledge that is pertinent to the Caribbean and to other regions that share the postcolonial experience as well as a similar dependency on tourism income.

But the realisation of our vision that was conceived during our postgraduate training in tourism has been achieved in concert and with the support of many persons to whom we wish to express our sincere thanks.

Specifically, we owe a debt of gratitude to our commissioning editor at Routledge for the opportunity to publish this work. We will never be able to repay in words of thanks the patience, support, and courtesy that were consistently administered to us as we painstakingly had to develop and hone the varied skills of time management, negotiation, and diplomacy that we have come to realise is the stock in trade of producing edited publications.

Thanks to our contributors who bought into this vision, made adjustments to their manuscripts, and who did not flag in commitment, but remained on board until the completion of the project.

We are grateful also to our families and to our friends for the assurance that they were always available whenever there was need for a listening ear, encouraging words at the other end of a telephone call, and in some instances, a warm hug.

Above all in an expression of our personal faith, thanks to God for the envisioning and enabling through to completion and also the hope for future endeavours.

Marcella, Donna, and Sherma,
July 2007

1 Caribbean Tourism

New Perspectives

Marcella Daye, Donna Chambers, and Sherma Roberts

INTRODUCTION

It has often been said, whether earnestly or jokingly, that Christopher Columbus was the first tourist to the Caribbean. Admittedly, it seems quite a dubious assertion that modern-day tourism in the region owes its genesis to the first landing of the European explorer. But from the vantage point of retrospection, it could be argued that, in some respects, tourism has been a major contributor to contemporary societal changes that are as indelible and irrevocable as those perpetuated on the landscape and natives by the peoples and civilisations that willingly came, or were forcibly transplanted, to the region. Today, there is an incessant flow of millions of 'arrivals' that visit the region mainly for the enjoyment of the sun, sand, sea, and sex that is epitomised in the Caribbean holiday experience. The consequences of the encounter and the accommodation between the peoples who now inhabit, and those who continue to come to the region as tourists, have determinedly contributed to the Caribbean being 'regarded as one of the world's premier travel destinations' (Duval 2004:3) and at the same time as being 'one of the most tourism dependent regions of the world' (Poon 2000:143; Jayawardena 2007:3).

Today, the economic prowess of tourism as a foreign exchange income earner as well as its contribution to the development of the Caribbean region are hardly contestable when examined from the lens of empirical economic indicators. According to Poon (2000:143), Caribbean tourism accounts 'for nearly one-third of all regional exports.' World Travel and Tourism (WTTC) statistics also indicate that whereas the Caribbean's tourism sector is ranked 13th in size globally, yet it currently ranks as number one in terms of relative contribution to regional economies (World Travel and Tourism Council [WTTC]:2007). However, the primacy of tourism in regional economies and societies encourages a tendency for the sector to be treated as a discrete entity that could possibly eclipse the sociopolitical and cultural context of the destinations where tourism activities takes place. This is often evidenced in discourses that relegate other interests for the sake of the tourism sector. But tourism for the most part occurs 'in situ,' which suggests that it undoubtedly shapes, and is also shaped by, the local context of its delivery. It is therefore the main intent of this volume to critique Caribbean tourism within the frame of the region's quest for modernity and self-determination. It seeks to foreground the varied ways and routes by which tourism activities relate and interact with Caribbean concepts that '[centres] in the historical experience of colonialism and the geosocial space of the periphery' as integral to the discourses of development in the region (Green 2001:41). According to Meeks and Lindahl (2001:xvii), the main currents of Caribbean critique are characterised by an attempt to rethink the political economy of the region in terms of the specific realities of Caribbean society. They suggest that the focal elements of Caribbean critique include themes such as conquest, colonisation, and postcolonial relations, the plantation economy and dependency, and the notions of indigenisation and creolisation. This volume therefore seeks to examine tourism development in the region in terms of some of these key themes with a view to facilitating the emergence of a Caribbean critique of tourism

In introducing this examination of Caribbean tourism, this chapter will first review the tourism performance indicators of the wider region, define the study area that is the focus for this book, discuss the purpose and objectives of this volume, and also provide a summary of the chapter contributions.

OVERVIEW OF TOURISM IN THE CARIBBEAN

The Caribbean archipelago comprises some 34 nation-states including islands located in the Caribbean Sea, the Atlantic Ocean, as well as countries on mainland South and Central America. The Caribbean Tourism Organisation (CTO) spatially defines the Caribbean as territories between the south of Florida in the United States, Cancun, in Mexico, Belize in Central America, Venezuela, Suriname, and Guyana in South America, as well as the islands of Bahamas and Bermuda located in the Atlantic Ocean.

With a population count of some 60 million, the region is a polyglot of English-, French-, Spanish-, and Dutch-speaking countries that all share the common history of European colonisation. Notwithstanding their spatial proximity and shared historical experiences, the countries of the region represent a differing array of size, population, culture and landscapes (Jayawardena 2002:88). Whether to a greater or lesser extent, most countries of the region are dependent on tourism as an engine of economic growth.

Tourism emerges as an attractive economic activity for the developing countries of the region because of its potential to foster GDP growth, to create employment, increase foreign exchange earnings, and attract capital investment. Up to the year 2004, the Caribbean attracted some 7 percent of world tourist arrivals. For that same year, tourism expenditures in the Caribbean increased by 7.9 percent, which represented earnings of US$21.6 billion. According to the World Travel and Tourism Council (WTTC), in 2005, tourism accounted for 15.4 percent of the region's gross domestic product (GDP), 15.1 percent of total employment, and 19.7 percent of total foreign exchange earnings. Forecasts from the World Travel and Tourism Council (WTTC) estimate that in 2007 travel and tourism in the Caribbean will generate US$56.1 billion of economic income, account for 16.5 percent of gross national product (GDP), as well as 14.8 percent of total employment in the region. The region's tourism sector has also been projected to grow by 3 percent in 2007, to be followed by a 3.3 percent annual growth between 2008 and 2017. However, concern has been voiced at the regional level on the comparative decline in the market share of the Caribbean. In an address on May 2007, Chairman of the Caribbean Tourism Organisation and Tourism Minister for St. Lucia, Alan Chastanet, noted that in spite of an international climate that favoured growth in the international travel market, the Caribbean's market performance was weakening.[1]

These tourism indicators, however, underscore the overdependence of many of the destinations on the fortunes of the industry. As open, dependent economies with a limited export focus, many of these countries rely on tourism as a lead sector to stimulate growth and development (McDavid and Ramajeesingh 2003:180). Although tourism is the fastest growing sector in the region, the high level of income leakage, ranging from between 25 percent in Antigua and Barbuda, 40 percent in Jamaica, and 56 percent in St. Lucia, means that a significant portion of earnings is lost to the region's domestic economies (ibid.). Furthermore, most economies are not able to maximise the benefits of their tourism earnings due to weak intersectoral links and limited economic diversification (Jayawardena and Ramajeesingh 2003:177). This suggests that, in spite of the success of the industry, the structural weaknesses of these economies persist. This limits their ability to harness the potential of the industry through economic linkages to primary sectors such as agriculture and agribusiness. The particular vulnerability of the region's tourism industry was underlined in the aftermath of the September 11 disaster in the United States. According to the Caribbean Tourism Organisation (CTO, 2007) Caribbean Tourism Statistical Report (2001–2005), tourist arrivals fell overall by 4 percent with a decline of 18.8 percent in the last four months in comparison with the previous year. The Caribbean relies heavily on the neighbouring North American mainland for tourist traffic, which is the primary market for the region. As Table 1.1 shows, the American market is the mainstay of the region's tourism industry. However, the United Kingdom and wider Europe still represent a potential growth market for the Caribbean.

Europe is placed as the second major market for the Caribbean, accounting for almost half of total arrivals compared to the total from the United States. However, European tourists are considered profitable, accounting

Table 1.1   Tourist Arrivals to the Caribbean by Main Markets

Country of Origin	2000	2001	2002	2003	2004
United States	10,469.4	10,239.2	10,132.3	10,665.3	11,443.9
Canada	1,211.1	1,256.4	1,273.1	1,500.6	1,739.1
Europe	5,180.0	4,825.5	4,492.9	5,010.3	5,289.6
Caribbean	1,285.9	1,208.6	1,302.3	1,423.6	1,429.1

Source: CTO—Caribbean Tourism Statistical Report 2004–2005

for a higher share of the total bed nights spent by tourists vacationing in the region since they usually stay longer than visitors from the United States (CTO 2007). Still, as long-haul destinations for Europe, the Caribbean faces the imperative of maximising their competitive position in the wake of increasing competition from areas such as the Asia-Pacific and the former Eastern bloc countries (Poon 2000:143). Intraregional competition is also heating up for Europe, particularly the UK market. In spite of their apparent natural advantage in the UK market, the English-speaking destinations of the region are facing fierce competition from Spanish-speaking countries such as the Dominican Republic and Cuba.

The Anglophone Caribbean

The specific focus of this book is on the English-speaking territories, excluding the US Virigin Islands, as a distinct grouping of countries within the Caribbean. These countries, also referred to as the Anglophone, Commonwealth Caribbean, or British West Indies, represent some eighteen nation-states. Today, the former British colonies in the region, as well as smaller territories currently under British rule known as overseas territories, still maintain some level of political and economic ties with the United Kingdom in spite of the dominance of the United States in the region.

As a subgroup in the region, countries of the Anglophone Caribbean experienced a slight decline of–2.8 percent in arrivals between 2004 and 2005, as shown in Table 1.2. This was mainly due to the decline in arrivals after Hurricane Ivan in 2004 that severely hit the Cayman Islands and Grenada and which led to dramatic falls in arrivals of–35.4 percent and–4.4 percent, respectively. The vulnerability of Caribbean destinations due to exogenous factors such as natural disasters is clear, but it is also likely that an overall attrition in arrivals to the region after Hurricane Ivan may have been stemmed by other destinations picking up a windfall from countries that were more severely affected by the hurricane.

Table 1.2 Tourist Arrivals in the Commonwealth Caribbean

Destination	2004	2005	% Change 2005/04
Commonwealth Caribbean	6,311.2	6,133.6	-2.8
Anguilla	54.0	62.1	15.0
Antigua & Barbuda	245.8	245.4	-0.2
Bahamas	1,561.3	1,608.2	3.0
Barbados	552.0	547.5	-0.8
Belize	230.8	236.6	2.5
Bermuda	271.6	269.6	-0.7
British Virgin Islands	p 304.5	p 337.1	10.7
Cayman Islands	259.9	167.8	-35.4
Dominica	80.1	79.3	-1.0
Grenada	133.9	98.5	-26.4
Guyana	122.0	116.6	-4.4
Jamaica	1,414.8	1,478.7	4.5
Montserrat	10.1	9.7	-4.4
St Kitts & Nevis	91.8	—	—
St Lucia	298.4	317.9	6.5
St Vincent & the Grenadines	86.7	95.5	10.1
Trinidad & Tobago	443.0	463.2	4.6
Turks & Caicos Islands	150.6	—	—

Source: CTO—Caribbean Tourism Statistical Report 2004–2005

According to CTO statistics, the slight drop of–2.8 percent in arrivals between 2004 and 2005 for the Anglophone Caribbean follows a similar decline in the Dutch West Indies of–0.3 percent and–63.7 percent for the French West Indian islands of Guadeloupe and Martinique. The U.S. territories of Puerto Rico and the U.S. Virgin Islands buck this trend with a 2.9 percent increase in the same period. However, the Spanish-speaking islands of the Dominican Republic and Cuba recorded arrival increases of 7.2 percent and 13.2 percent, respectively. This discussion of arrival indicators demonstrates that destinations within the Anglophone Caribbean show a comparable rate of growth. Intraregionally, these statistics also show, as mentioned earlier in this discussion, that Cuba and the Dominican Republic are strong performers, and, apart from Cancun and Cozumel in Mexico, are the main competitors to the Anglophone Caribbean in the region.

PURPOSE OF THE BOOK

In spite of recent indicators pointing to a slowdown in the rate of growth, the relatively buoyant and stable economic performance of the tourism sector in the Anglophone Caribbean may have contributed to the entrenchment of views that affirm tourism development as an imperative for most countries. For those detractors who take on board the attendant negative impacts of the tourism industry, the view is less bullish, and tourism is seen perhaps as a 'necessary evil' (Puttullo 2005). Polar enquiries into tourism as a 'blessing or blight' (as first described by George Young)[2] that mainly orient tourism research seem limited in relation to a region that is already committed to, and is increasing the level of, participation as key players in international tourism. However, normative perspectives of Caribbean tourism development, while valid and beneficial, do not usually take into account the 'situatedness' and the particular confluence of history, culture, political, and economic aspirations of Caribbean postcolonial societies in the quest for 'competitive ascent' in the global community (Northover and Crichlow 2007:207) .

As a pervasive, dominant, and ever-increasing sphere of activity, there is a need for the production of knowledge on Caribbean tourism that addresses the specific realities that are unique to Caribbean countries that may not be readily served by traditional and existing models and concepts that have been successful in other locations. The Caribbean bears similarities but yet has remarkable, distinctive features of historical and political evolution, mixtures of races and cultures as well as physical landscape attributes (Girvan 2001:4) . But beyond the adaptation of models and concepts of development that suit the particular demands of Caribbean tourism, there is also the need to address and redress normative views that limit and restrict new thoughts and perspectives of tourism, in the drive for equitable and lasting individual country and regional development.

Regional scholars have, pre- and postindependence, been involved in reconstructing and producing 'indigenous interpretations' of the Caribbean experience, and, according to Caribbean historian Verene Shepherd (2007:2), the rationale of Caribbean revisionism of imperial historiography and perspectives has been to challenge existing knowledge that was purposed to 'exercise power and authority' over colonial subjects. The generation of knowledge and critiques of Caribbean societies has been motivated by the quest for freedom and power, which Nettleford (1995:82) defines as the '[capability] to make definitions about oneself and to follow through with action on the basis of those definitions . . .' This can be related to efforts to manage and plan for the tourism industry's current needs and those of future generations, which has implications for agency and the extent of control that regional countries have in charting desirable futures. However, this endeavour may involve not only probing but, perhaps, also challenging normative and prescriptive views of Caribbean tourism.

This edited volume therefore seeks to expand the scope of the interrogation of tourism in the Anglophone countries of the region to incorporate ideas and perspectives from contemporary Caribbean thought that are largely not represented in existing epistemologies of tourism in these destinations. Accordingly, this text explores tourism in the Caribbean in relation to colonisation and postcolonial relations, the region's geopolitical formations and conjunctures, as well as issues of culture, race, identity, and governance. The chapters presented in this volume reflect the intent of this book to engage plural discourses and disciplines in an examination of the processes and development of tourism so as to build on and deepen analytical repertoires and frameworks to clarify the unique experiences and practices of Caribbean tourism.

Conceptualising the Caribbean

It has been said of the region that 'there are many Caribbeans'(Girvan 2001:3). This may seem to be a fitting description based on the diversity of cultures and sociopolitical formations of regional groups and subgroupings that are represented across the region. These are features of European colonial legacy that have resulted in nation-states that are separated by language, political systems, and a tradition of insularity and noncooperation between countries. Even for Anglophones, attempts at transcending inherited colonial demarcations have been uneven, albeit with some success in economic union among the islands of the Eastern Caribbean, and the formation of regional bodies but notable failures in efforts at wider political federation.

Ironically, while the colonial enterprise may have fractured the region, this shared historical past has also been the basis for a common identity (Dash 2003:287). According to Girvan (2001:4), the Caribbean has been reinterpreted and reinvented by 'native scholars as expressions of intellectual and political resistance' in 'response to external influences and internal currents' (Girvan 2001:6). Relating this to the tourism industry, an example of reinvention is demonstrated by the Mexican provinces of Cancun and Cozumel being marketed as the Caribbean even though they also assume the designation of Latin American. Moreover, Girvan (2001: 6) points out that the notion of Caribbean has also become transnational with a growing interest in studies on the Caribbean diaspora communities.

The process of the construction of the Caribbean has therefore been integral to the drive towards the development and empowerment of the region. The main thrust of a Caribbean conceptualisation of development has been to counter Western paradigms that do not take into account 'the messy historical realities of colonialism, dependency and continuously reproduced heterogeneity and hybridity' of Caribbean societies (Green 2001:41). Consequent development of Caribbean thought has therefore centred the historical, social, and cultural evolution of these societies, creating theories and concepts such as 'creolisation, ethnogenesis, inter-ethnic fusion, hybridisation, merger, transculturation and pluralism' (Shepherd and Richards 2002: xi). These theories mainly seek to clarify the dynamic changes and outcomes of the encounters between the many ethnic groups and cultures in the region. A similar project has also been ongoing in the fields of political economy, governance, sociology, and gender studies in the Caribbean spawning many indigenised theories and interpretative models (Green 2001:40–41). However, cultural interpretations and concepts, mainly creolisation, have been particularly dynamic in scope and influence both locally and globally and have been influential in many different fields and disciplines. According to Shepherd and Richards (2002:xii), creolisation is the 'dominant intellectual construct,' particularly in the fields of history and cultural studies, and is progenitor of theories such as Mary Louise Pratt's transculturation and Homi Bhabha's hybridity (Shepherd and Richards, 2002:xiv). Creolisation is antiessentialist, originally conceptualised as a subaltern strategy of resistance by means of mixing and syncretising varied cultural forms to create new expressions that are adaptable and transcendent. It constructs Caribbean societies as subversive in resisting normative discourses and praxis of 'Othering' with the intent of creating space for subaltern voice and agency (Sheller 2003:195). Yet, creolisation is not without contestations that view its limitations in explaining the complexity of power relations between ethnic groups, class, and political elites in individual countries and across the Anglophone Caribbean. Thus Khan (2004:167) argues that the state-led nationalist representations of the multiethnic Caribbean island Trinidad as a callaloo ('a multiple-ingredient stew, also the national dish'; ibid.,174) nation is an appropriation of the concept of creolisation that Indo-Trinidadian nationalist discourse views as a 'subvention within a vision not its own.' Dash (2003:297) points to the problematique of political identity in creolisation discourses that predisposes a tendency towards 'exclusionary constructs' of identity. While in their review of the political economy of the Caribbean, Northover and Crichlow (2007:223) contend that more research is required into the process of creolisation to probe the complexity of social power that relates to the continued poverty and exclusion of marginalized groups in Caribbean societies.

Interestingly, whether in support of or opposition to (Shepherd and Richards 2002:xiv) creolisation, it is often a dominant referent in the articulation of a Caribbean ontology (Henke 1997; Northover and Crichlow 2007). This may be mainly due to its explicatory power of the complexities and possibilities of Caribbean societies. Northover and Critchlow's (2007:223) quote of Lloyd Best's commentary on the complexity of Caribbean ontology is worth repeating here—'emergent from unique yet systemic global and local,' 'structural and institutional', 'plantation and postcolonial,' 'processes and forces.' This edited volume utilises this ontology of the Caribbean in the engagement of new perspectives on the development discourses and practices of tourism in the region. Hence, the word *new* is operationalised in this work to represent the adoption of this Caribbean ontology as a frame for interrogating and constructing knowledge on tourism in the region.

Critiquing Caribbean Tourism

In pursuing this examination of Caribbean tourism, this volume is divided into two main thematic strands: image, culture, and identity and governance. The first part, image, culture and identity, opens with an interrogation of stereotypical representations of Anglophone destinations and the ways in which these images tend to prescribe and impose ways of seeing the region that may be contradictory to the developmental goals and aspirations of these destinations. In this chapter, Daye contends that the reliance of Anglophone Caribbean destinations on the stereotypical appeal of sun, sand, and sea may be limiting the competitiveness of the region in the global marketplace. Daye argues that images of Paradise have been rooted in hegemonic discourses that reduces the particularity of destinations; and, in turn, introduces the Caribbean critique of creolisation as a differentiation strategy for branding the region. While Daye mainly presents creolisation as a seamless process with generic relevance and application to the Anglophone Caribbean, Ramcharitar in Chapter 5 challenges the homogenising assumptions of this concept in his examination of the cultural fusion of nationalism and tourism in the staging of festivals and cultural activities in the islands of Trinidad and Tobago. As an Indo- Trinidadian, Ramcharitar views the emphasis on ethnicity and the 'folk' in the branding of the tourism destination as a carefully managed project of a nationalist exclusionary discourse that is limiting the 'possibilities of cultural fusions and interactions of European, Asian, and African cultures' in Trinidad and Tobago. Reddock (2002:119) has pointed out that such contestations on the 'content and definition of "national culture" in Trinidad and Tobago' have been a significant preoccupation in the republic and are likely to continue for some time to come. Ramcharitar's chapter demonstrates the extension of this contestation to yet another fairground for this struggle on national identity, namely, tourism. As suggested by Reddock (2002:126), there is a need for more research that will offer insightful critical understandings of the processes of contestation and interculturation of various ethnic and social groups in the formation of the political culture of Caribbean societies.

If the relationship between tourism and the 'dynamics of relational identity' of Caribbean societies is fractious among ethnic groups at the country level, it is no less contentious and controversial at the international level. In Chapter 6, Chambers' foray into the prickly ground of the attitudes of Jamaicans to homosexuality and, by extension, gay tourism underscores the implications for the brand image of the destination. Chambers contends that the antigay stance of Jamaicans may be seen as an assertion of a postcolonial identity, particularly in the face of external threats to boycott local hotels that prohibit homosexual couples. Here Chambers argues that by depicting the prevailing negative attitudes of Jamaicans to the homosexual lifestyle as barbaric and uncivilised, former colonies such as Jamaica are being constructed as deserving of approbation and sanctions in order to ensure conformity to the more enlightened worldview of developed, capitalist societies. However, Chambers postulates that some level of adaptation is perhaps likely considering the reliance of the Jamaican economy on the profitability of the tourism sector and the lucrative nature of the gay market within this context. The question for further study would be the nature of this accommodation, whether it would be mainly a demonstration of token acceptance tolerated only within the tourist space or a more radical shift in deep-seated antipathy to the homosexual lifestyle.

Conversely, Jamaica's cultural identity in the form of reggae music is globally celebrated and even venerated. The iconic popularity and acclaim of Bob Marley have been officially incorporated as part of the promotional strategy for Jamaica's destination image. In their discussion of the universal appeal of Bob Marley as a global symbol of resistance and also of 'human hope,' Jalani and Sonjah Niaah in Chapter 3 explore the development of the Bob Marley Museum as the main cultural tourism attraction and as a site of pilgrimage in Jamaica. Their examination of the comments of tourists who visited the museum uncovered an appreciation and association of Western tourists to the themes of dissent, liberation, and universal equality trumpeted in the music of Bob Marley. Notably, Caribbean diaspora visitors who trace their roots to the island or the other Caribbean countries viewed these sites not just as touristic attractions but as a homecoming and an affirmation and sense of pride in their Black identity. The authors also argue that the emergence of a 'reggae aesthetic' as an expression of identity and protest has operated as a fillip for the island's tourism promoters to the extent that it accommodates the negative images of the tumult and the tensions of local disturbances as part of the island's 'appeal.' The manifestations of reggae tourism in this chapter are therefore traced to the popular and well traversed routes of the Bob Marley museum, the 'tenement yard in Trench Town,' where he lived during his formative years and his burial shrine.

But Webster in Chapter 4 uncovers and probes a novel and 'off the beaten track' mode of cultural appropriation and consumption of Jamaica's musical heritage and resources in the form of those who visit the island for the purpose of purchasing original reggae vinyl records. As a musicologist, Webster's insider knowledge on the activities of this unique expression of reggae tourism presents a fascinating exploration of the appeal of reggae in attracting visitors to Jamaica. Webster's methodology in tracking this underground culture is that of a participant observer who engages in the ethnography of this phenomenon. His interviews with vinyl tourists from

Europe, Japan, and the United States show their indifference to the ubiquitous sun, sand, and sea image of the island, typified in their common assertion of never going to a beach on the island or, in the case of one informant, going only once. These are quintessential music tourists whom Gibson and Connell (2005:171) describe as people who are '[more] willing to travel to particular destinations because of music.' However, Webster's exposure of this niche within a niche is instructive in demonstrating the high currency and valorisation of what is constructed in the mind of these tourists as being culturally authentic. It raises questions as to the extent their activity may be considered to be cultural piracy if the value these vinyl tourists ascribe to the provenance is not accorded similar worth by those who supply the trade. This variant of cultural tourism may not attract the attention of local tourism promoters or cultural brokers by reason of its speciality and few numbers, but it illustrates the idiosyncratic and selective permutations of cultural consumption in motivating and stimulating travel (Gibson and Connell 2005:92). The search and forage expeditions of vinyl tourists in the rural hinterlands, the back streets of downtown Kingston, and uptown and middle-class suburbia in Jamaica epitomise a tourist quest for authenticity that is prescriptive and measurable in terms of motivation and activities.

But as Dodman and Rhiney in Chapter 7 point out, authenticity in the presentation of local food often bears no direct connection to provenance but is usually determined by a construction of the exotic to satisfy notions of difference or something unlike the everyday. Additionally, cuisine in the tourism experience has been linked to notions of cultural capital and cultural appropriation that are enacted in the consumption of food. In their exploration of authenticity of local cuisine as part of the holiday experience in Jamaica, Dodman and Rhiney link the local agricultural supply of food to the provision of local cuisine in the hotel sector and at local restaurants. They identify tourist demand, the nationality of the chefs working in the main hotels, and access to and supply of local agricultural products as key variables in determining the availability of local cuisine for tourists. Inevitably, notions of authenticity are complicated by the reality that local cuisine is also creolised, bearing the features of the admixtures of local cultures and ethnicity as well as the adaptations to the sensitivities of the Western tourist palate. Dodman and Rhiney therefore conclude that local cuisine offers some prospects for enriching the tourist experience and attracting new markets.

The chapters in this section on Caribbean image, culture, and identity all contribute to understandings of the ways in which tourism development may be challenging the values that underpin the ways Caribbean identity and culture are constructed and negotiated in an increasingly globalising and homogenising world. In many ways the packaging and presentation of tourism in these countries involve a process characterised by tensions, concord, and adaptation between locals and tourists that is dynamic and evolving.

However, a limitation of this section is that many of the examples presented are based on Jamaica. This should not be seen as an inference that these examples of Jamaica typify the entire region or other individual countries. The focus on Jamaica is mainly an artefact of the direction of the research rather than an assumption of a 'Jamaicanisation' of the Anglophone Caribbean. In some respects the case study examples represent a starting point for various explorations that may be extended to other regions where such considerations are relevant.

Part 2, on governance, presents an eclectic range of studies that demonstrate the preeminence of government policy and decision making in Caribbean tourism. The chapters in this section acknowledge that in the quest to harness the potential of the industry for wider development goals, governance has become a critical focus of deliberation and activity in tourism development for Caribbean destinations. In this context, topical issues of the sustainability of tourism in the region, social exclusion and poverty alleviation, tourism education and training, political and regional partnerships are reviewed and interrogated. Throughout this section, the underlying contention is that templates or 'best practice' case studies from without the region may not generally offer viable solutions for the peculiar problems and challenges of tourism development in Caribbean societies. Therefore, the notion of emergent management approaches and the processes of adaptation and indigenisation are implicit in the studies presented in this section.

In Chapter 8, Roberts begins this exploration of governance and Caribbean tourism by arguing that metanarratives on sustainable tourism development that have their origins in developed countries do not necessarily account for the lived realities of local communities in developing countries. This chapter suggests that a critical understanding of sustainable tourism development in developing countries is required to clarify the ways in which the governance structures and contextual realities have shaped the understanding and praxis of sustainable tourism development. Here, Roberts makes the case for the examination of the decisions and operations of local small businesses in Tobago in order to derive insights on the operationalisation of the concepts of sustainability in small Caribbean island states. Indeed, Reddock (2002:126) has pointed to the importance of examining social phenomena in Caribbean societies by investigating the 'little tradition,' that is, people who seem to be of low status and influence. In Chapter 9, Brooks spotlights the 'little tradition' by focusing on the squatter communities located in the vicinity of the tourist resort town of Ocho Rios in Jamaica. Brooks's investigation shows a pattern of sociospatial exclusion of these marginalized communities and contends that the development of the tourist resort town of Ocho Rios privileges particular tourism interests while excluding poorer communities. Interviews with informants from these communities indicate their worldview in which they are positioned as actors in the tourism sector even though their informal trading activities and presence in the resort areas are prohibited. Their entrepreneurship and enterprise, although engaged at the informal level, are not acknowledged as a potential to be harnessed, but is usually constructed as threatening tourism in the town. Brooks's work is informative in terms of its method-ological detail on sampling and data collection and also in exposing the politics of exclusion in the discourse and praxis of tourism planning and development in the resort town. Northover and Crichlow (2007:211) have suggested that in the quest for modernisation, Caribbean societies need to consider the potential for the emergence of 'an emancipatory creole social power.' From a critical realist stance, they posit that the processes of gov-ernance in Caribbean societies are implicated in the creation of noncitizens or marginalized 'Others.'

The challenges of governance are further illustrated by Grenade's review of HIV/AIDS in Chapter 10. In a case study of the island of Gre-nada, this chapter probes the problem of policy planning and implemen-tation in terms of the island's tourism dependency and the threat which HIV/AIDS poses for development. Grenade details the various levels of interactions that occasion the policy planning for this threat and the reali-ties of 'power differentials between tourists and locals, the 'hedonistic nature of tourism; as well as the outcomes of the structural adjustment programmes that have weakened institutional capacity in the country. Grenade's examination exposes the deficiencies in governance manifested in disjointed or absent policies and foregrounds the need for creativity and community engagement as well as policy commitment and imple-mentation. Jordan's discussion in Chapter 11 on regional partnerships as a strategy for tourism development in the Anglophone Caribbean shifts the focus of governance from individual countries to the regional level. In spite of a chequered history of both successes and failures in regional cooperation, Jordan maintains that such institutional cooperation is worthy of consideration given the exigencies of the dependency and vul-nerability of small island states in the Caribbean. Jordan elucidates the benefits and challenges of partnerships at the governmental and organisa-tional level, while contending that such arrangements are as important as the pursuit of extra regional links.

In Chapter 12, Lewis continues the theme of regional cooperation by pointing to the benefits of a common approach in building development models that suit the needs of Caribbean tourism specifically in the area of tourism and hospitality training. Lewis argues that new tourism knowledge is required as 'Western' models have been imported without taking suffi-cient account of the needs of the local tourism sector and the existing social, cultural, and economic framework of the host country.'[3] In redressing this deficiency, Lewis advocates a stakeholder informed approach in designing the content and delivery of tourism education in the region so as to institute programmes that deliver not only efficiencies in craft and vocational skills but foster competencies in managerial performance, leadership, innovation, and creativity. This regional training agenda would also be conducive to the creation of pedagogies that both equip and empower Caribbean nationals in the quest for modernity and self-determination.

The chapters in this volume thus represent important issues for the tourism sector in the Caribbean, but they also undoubtedly reflect the main research interests of the contributors. There are notable omissions in this volume of critical areas that were not covered directly, such as gender, class, sport tourism, the cruise and hotel sectors, and transport, among others. There was also an underrepresentation in the case studies of some of the countries in the Anglophone Caribbean, but this has also been a consequence of the dearth of specific research on some individual countries. It is hoped that this volume exposes these gaps and in turn encourages a wider range of coverage of research on individual islands and issues relating to Caribbean tourism.

The search for pathways of development that are applicable to the needs and context of the Caribbean inevitably involves the engagement of multiple viewpoints that consider global or 'etic' perspectives that have to date dominated studies on Caribbean tourism; as well as the 'emic,' that is, the insider view. In bringing to the fore the emic perspectives, the contributors of this volume are all Caribbean nationals who have attended and/or currently teach at the University of the West Indies. This emic approach explains the title of the volume *New Perspectives in Caribbean Tourism;* the use of the proposition *in* demarcates the insider views presented in the chapters. At the onset, however, it is important to acknowledge previous work that has been done in this tradition that has initiated this current focus. From as early as 1973, Bryden (1973), in his book *Tourism and Development: A Case Study of the Commonwealth Caribbean,* pioneered this enquiry in his questioning of tourism's role as a tool of economic development for the region. However, this review was mainly restricted to economic development and did not focus on other areas of tourism's pervasive influence on the society and culture of the region. Twenty years later, in the 90s, Palmer (1994) challenged the perpetuation of the relationship between tourism and colonialism in terms of the implications for national identity in the Bahamas. She maintained that stereotypical representations of the region that emphasised the colonial era instead of postindependent achievements threatened the ability of Bahamians to progress from the limitations of their colonial identity (Palmer 1994:806). Recent publications by (Sheller 2003) and (Thompson 2006) have traced the evolution of hegemonic discourses in inscribing economic relations and the tropicalisation and imaging of Caribbean destinations. The authors mentioned here are some key and current examples. They are by no means exhaustive of the varied studies and investigations that have engaged in a Caribbean critique of tourism in the region. However, it is hoped that this volume contributes to this area of Caribbean studies and tourism scholarship and, more importantly, encourages this line of research to be systematically sustained and pursued even as it seems likely that the tourism industry in the region will maintain its primacy and influence in Caribbean societies.

NOTES

1. Noted in an article published in the *Daily Gleaner* on May 17, 2007, entitled 'Caribbean Tourism.'
2. George Young's seminal work on the impacts of mass tourism, *Tourism: Blessing or Blight,* was published in 1973.
3. Lewis attributes this quote to an article by H. L Theuns and F. Go in 1992 entitled 'Need Led" Priorities in Hospitality Education for the Third World' published in *World Travel and Tourism Review* 2:293–302.

REFERENCES

Bryden, J. M. 1973. *Tourism and development: A case study of the commonwealth Caribbean.* London: Cambridge University Press.

Caribbean Tourism Organisation (CTO; 2007). Caribbean Tourism Statistical Report 2001–2005. St. Michael, Barbados: Caribbean Tourism Organisation.

Dash, J. M. 2003. Anxious insularity: Identity politics and creolization in the Caribbean. In *A pepper-pot of cultures: Aspects of creolization in the Caribbean,* G. Collier and U. Fleischmann, eds. Amsterdam and New York: Editions Rodopi.

Duval, D. T. 2004. Trends and circumstances in Caribbean tourism. In *Tourism in the Caribbean, trends, development, prospects,* D. T. Duval, ed. London and New York: Routledge.

Gibson, C., and J. Connell. 2005. *Music and tourism: On the road again.* Clevedon, UK: Channel View Publications.

Girvan, N. 2001. Reinterpreting the Caribbean. In *New Caribbean Thought: A reader,* B. Meeks and F. Lindahl, eds. Kingston, Jamaica: University of the West Indies Press.

Green, C. 2001. Caribbean dependency theory of the 1970s: A historical-materialist-feminist revision. In *New Caribbean thought,* B. Meeks and F. Lindahl, eds. Kingston, Jamaica: University of the West Indies Press.

Henke, H. 1997. Towards an ontology of Caribbean existence. *Social Epistemology* 11 (1):39–58.

Jayawardena, C. 2007. Caribbean tourism: For today and tomorrow. In *Caribbean tourism: More than sun, sand and sea,* C. Jayawardena, ed. Kingston, Jamaica: Ian Randle Publishers.

———. 2002. Mastering Caribbean tourism. *Journal of Contemporary Hospitality Management* 14 (2):88–93.

Jayawardena, C., and D. Ramajeesingh. 2003. Performance of tourism analysis: A Caribbean perspective. *International Journal of Contemporary Hospitality Management* 15:176–79.

Khan, A. 2004. Sacred subversions? Syncretic Creoles, the Indo-Caribbean, and "Culture's in-between." *Radical History Review* 1 (89):165–84.

McDavid, H., and D. Ramajeesingh. 2003. The state and tourism: A Caribbean perspective. *International Journal of Contemporary Hospitality Management* 15:180–83.

Meeks, B., and F. Lindahl, eds. 2001. *New Caribbean thought: A reader.* Kingston, Jamaica: University of the West Indies Press.

Nettleford, R. 1995. *Inward stretch outward reach: A voice from the Caribbean.* New York: Caribbean Disapora Press Inc.

Northover, P., and M. Crichlow. 2007. Freedom, possibility and ontology: Rethinking the problem of 'competitive ascent' in the Caribbean. In *Contributions to social ontology*, C. Lawson, J. Latsis, and N. Martins, eds. London and New York: Routledge.

Palmer, C. 1994. Tourism and colonialism: The experience of the Bahamas. *Annals of Tourism Research* 21 (4):791–811.

Poon, A. 2000. The Caribbean. In *Tourism and hospitality in the 21st century*, A. Lockwood and S. Medlik, eds. Oxford: Butterworth Heinemann.

Puttullo, P. 2005. *Last resorts: The cost of tourism in the Caribbean*. London: Latin America Bureau.

Reddock, R. 2002. Contestations over culture, class, gender and identity in Trinidad and Tobago. In *Questioning Creole: Creolising discourses in Caribbean culture*, V. Shepherd and G. Richards, eds. Kingston, Jamaica: Ian Randle Publishers.

Sheller, M. 2003. *Consuming the Caribbean: From Arawaks to Zombies*. London and New York: Routledge.

Shepherd, V. A. 2007. *I want to disturb my neighbour*. Kingston, Jamaica: Ian Randle Publishers.

Shepherd, V. A., and G. L. Richards, eds. 2002. *Questioning Creole: Creolisation discourses in Caribbean culture*. Kingston, Jamaica: Ian Randle Publishers.

Thompson, K. A. 2006. *An eye for the tropics: Tourism, photography, and framing the Caribbean picturesque*. Durham, NC, and London: Duke University Press.

World Travel and Tourism Council (WTTC), World Travel and Tourism Council, Year 2007, Tourism Satellite Accounting Research (Caribbean), Online. http://www.wttc.travel/eng/WTTC_Research/Tourism_Satellite_Accounting/TSA_Regional_Reports/Caribbean/index.php. (Date accessed: 22 Feb 2007).

Part I

Image, Culture, and Identity

2   Re-visioning Caribbean Tourism

Marcella Daye

Just mention the Caribbean and eyes light up with excitement. The name conjures up images of warm sandy beaches, turquoise waters and palm trees, rum punch, smiling happy faces and the sound of steel drums in the distance. 'Caribbean, everything you want it to be,' CTO TRAVEL FAX

We in the Caribbean have not built pyramids, pillars, cathedrals, amphitheatres, opera houses etc that are the wonders of the world, but we have more creative artists per square inch than is probably good for us. In addition we have created and are creating mental structures which are intended to be the basis for that self-confidence, that sense of place, purpose and power without which there can be no integration of inner and outer space. (Nettleford 1995:83)

INTRODUCTION

The purpose of this chapter is to interrogate stereotypical images of the Caribbean holiday experience in terms of their effectiveness in enhancing the region's global market positioning and competitiveness. By critiquing the various meanings and contestations of conventional images of the region as sites of Paradise, this chapter contends that these image constructs may be restricting opportunities to build differentiated, value-added images of the Caribbean brand. By introducing the Caribbean critique of creolisation as a differentiation strategy for the region, this chapter seeks to open up dialogue on ways of negotiating and resisting normalising and homogenising discourses, in order to build unique selling propositions and identities of the Caribbean brand in the global tourism marketplace.

The Caribbean Paradise Archetype

The Caribbean is often universally invoked as a signifier of sun, sand, and sea hedonistic holiday experiences. The stability of these images, from the incipient years of tourism in the region to contemporary times, could logically be considered as a boon for regional promoters in marketing their destinations. Destination advertising and promotional media collateral of the Caribbean abound with stylised images of coconut-fringed beaches, fecund vegetation, riotously bright-coloured costumed island revellers, smiling hosts, and suntanned, sated guests. These stock images are continually reproduced and disseminated, maintaining their currency as the standard 'tried and true' grist of the imagery mill of Caribbean tourism advertising.

Apart from a slight decline in arrivals in 2001 and 2002, statistics from the Caribbean Tourism Organisation (CTO) show steady growth in tourist arrivals to the wider Caribbean region. Total tourist arrivals from the United States to the Caribbean (inclusive of the English-, Dutch-, French-, and Spanish-speaking territories) show a 20.5% increase between 1999 and 2004. For the same period, there was a 61.7% increase in arrivals from Canada and a 5.3% increase from Europe.[1] The steady growth in tourist arrivals attests to the region's success in purveying and attracting visitors to the region using their stereotypical representations of the destinations.

Beyond the general trend of positive tourism-performance indicators, there is also the apparent affirmation of the Caribbean's performance in the international tourism arena with many destinations, local hotels, and tourist boards winning accolades and awards from the travel trade and travel magazines' consumer polls as the 'Best' in the world. Although these awards may be challenged or even discounted as artefacts of public relations and political lobbying within the travel trade, nevertheless they indicate the growing prestige and leverage of the Caribbean in global tourism.[2] More significantly, however, they also function to legitimise the existing tropes and methods of imaging and marketing the Caribbean to the world.

Even among those whose vocation it is to engage reflexively and question established views, these stereotypes may also be entrenched. In a review of Mimi Sheller's 2003 *Consuming the Caribbean,* David Lambert (2004:138) stated:

> If you tell someone, even a fellow academic, that your research interests lie in the Caribbean, you are likely to encounter smirks and smiles—perhaps envious, conspiratorial or slightly scornful; 'that must be fun' is common reply. Such responses betray deep-seated preconceptions about the Caribbean as a place of sensuous enjoyment, pleasure and relaxation, a place where rigorous intellectual work, is for some, difficult to envisage.[3]

However, it would not be fair to infer that this quote typifies a catholic, uncritical perception of the Caribbean. But it may be argued that it clearly demonstrates the dominance of touristic images in the epistemology of the region. According to Therkelsen (2003:135), such stereotypical images are valuable props that facilitate effective marketing communications where a priori cultural images are the practical basis for fashioning the product to meet the needs and demands of the target market. Therkelsen (2003:135) explains that 'the rationale behind this is that tourism destinations seem to generate certain internationally shared meanings which may function as common denominators for an international promotion strategy.'

Therkelsen (2003:135) points out that for the tourism marketer, there is substantial merit in reproducing stereotypes and clichéd images as they reflect 'consumers' standardised needs and frame of mind' particularly in the tourist quest for the extraordinary, and to get away from everyday life.' Loomba (1998:59)suggests that stereotyping is a method of information processing that involves the reduction of images and ideas to more manageable forms. Within the context of the international tourism industry, the global span and appeal of these stereotypes function to bridge the forces of demand and supply. As such, destinations are networked and connected in an international web of markets and a 'system' that generates the flow of tourist traffic around the world. As a region boasting tropical attributes conducive to all-year-round vacationing, the Caribbean has been mapped out in this transnational design as sites of 'Paradise' equipped to service market demand for diversionary experiences. The stereotypical dominance of these images gives credence to and focuses marketing activities that are designed to stimulate consumer travel. The apparent success of this strategy in terms of results of growing tourist traffic leads to the assertion that since the region is firmly established as an archipelago of fantasy and desire, then it may be merely contentious to interrogate these marketing images.

The axiomatic power of representations of the Caribbean as images of Paradise may disguise the sophisticated machinery and craft that sustain these images. Marketing praxis essentially involves the habitual reinforcement of discourses that attempt to influence consumer behaviour (Moisander and Valtonen 2006:10). Stereotypes appear normative through a continuous marketing process of reaffirmation, repackaging, and updating for consumers where cultural myths and traditional symbols are invoked to shape and direct consumer's perceptions and attitudes(Arnould and Thompson 2005:875). Effective marketing is therefore a process that both mirrors and shapes consumer preferences and trends in demand.

Stereotypes of Caribbean Paradise may seem inviolable within the disciplined framework of the 'circuit of culture,' but this may belie moments of contestation and challenge that are usually erased or silenced in this discourse. According to Moisander and Valtonen (2006:7), today's marketplace is a 'joint cultural production of marketers and consumers where meanings, values and ideologies are constantly being negotiated.' This implies the possibility of consumers resisting stereotypical images and also marketers attempting to create and introduce new trends in the marketplace. Those who study consumer culture assert that meanings that inform consumptive practices tend to change and respond to shifts in wider cultural tensions and sociocultural movements (Arnould and Thompson 2006:876). The point of focus is that meanings that inform consumptive practices are not static. This suggests that even unassailable stereotypes evolve and may be disrupted.

However, threats to the stability of images of Caribbean Paradise may not always result from endogenous market dynamics, but at times they may be an outcome of exogenous factors. Images of Paradise in the region often require high marketing spend during an active hurricane season, incidents of civil unrest, or protracted media coverage on various destinations of drug trafficking or missing American teenagers.[4] After decades of experience in managing the vagaries of such events, regional tourism marketers are by now well-practised and rehearsed in producing arsenals of effective promotional and advertising campaigns and media messages of 'Paradise Restored' and 'back in business.'

Nevertheless, these instances of vulnerability, to which the regional destinations are often individually and collectively exposed, tend to bring into sharp relief the substitutability of countries of the Caribbean. The easy transferability between sites of Paradise demonstrates clearly there is no inherent fixity of these images to a specific destination or place. Consumers are not necessarily as loyal to destinations as they are to their inner motivations to escape to sites of pleasure. For Featherstone (2005:214), this simple transferability relates to the postmodern condition that tends to decouple countries and sites from their cultural and social contexts. This results in the emergence of so-called 'nonplaces' that are created and experienced in terms of their exchange function of meeting consumer needs. Featherstone suggests that airport terminals, motorways, service shops, shopping malls, and tourist beaches are all examples of nonplaces in contemporary, postmodern societies.

> The tourist beach can be seen as another such non-place. Stripped of geographical and political specificity and fixed within its elements of white sand, blue sea and clear skies, the tropical beach can be endlessly reconfigured as a desirable tourist destination that is almost independent of social context. Remove the signifiers 'Goa,' 'Cuba,' 'Gambia' or 'Barbados' from photographs in tourist brochures, and the resorts become indistinguishable places arranged for tourist pleasure, and reached as an end point of a journey through the circuits of terminals, transit areas and transport systems that process the consumption of the experience. (Featherstone 2005:214).

From this vantage point, it is clear that the legacy of history and culture, wealth of resources, and the creativity of the peoples of countries that are endowed with tropical attributes are not necessarily privileged in the Western imagination of Paradise. This may seem to be a travesty for some destination promoters who spend millions in building distinctive brands for competitive advantage in the marketplace. It is apparent that the underlying discourse of Paradise is interlocked within a marketing discourse of substitutability, one that empowers the consumer to transpose this ideal easily to whatever places and locations that can accommodate and replicate these attributes and values. Essentially, this suits the transnational purveyors of packaged holidays who produce mass, undifferentiated products because it not only ensures a catalogue of ready markets globally, but it also facilitates accessibility and lower costs for consumers in main tourist-generating countries. Destinations therefore may not have much room to negotiate or contest constructions of Paradise, if they are constrained to conform and package their landscape at the behest of the consumer market.

This preceding discussion sets the background for the objective of this chapter, which is to examine the imaginary of the Anglo Caribbean holiday experience. Essentially, this chapter probes the nature of contestations and negotiations of current marketing images of the region and the extent to which there are specified, denotative meanings of 'Caribbean-ness' that presents a distinct brand image and identity for destinations in the Anglophone Caribbean (Edmondson 1999:2). It also seeks to explore whether there is evidence of interventions or disruptions in the hegemonic discourses of the region in the self-representations of Caribbean destinations in their tourism promotional material. This enquiry is conducted by discursive content analysis and critiques of newspaper abstracts selected from travel articles in *The Observer* and the *Sunday Times* in the United Kingdom, as well as the promotional information presented on the official Web site home pages of the countries of the Anglo Caribbean.

INTERROGATING PARADISE

Counterdiscourses to hegemonic representations of the Caribbean have traditionally focused on the reductionism inherent in the imaging of the region as liminal spaces for the Western construction of Paradise (Palmer 1994; Sheller 2003; Puttullo 2005). While idealistic, images of fantasy and paradise continue to resonate with Western tourists, it may also be argued that these same images work in the reverse to trivialise or reinforce notions of the region as merely 'playgrounds' and 'peripheries of pleasure.' In a critical review of the marketing and advertising of Hawaii, Costa (1998:338) contends that paradisiacal discourses and tropes are not beneficial because they function as instruments of control that essentially prescribe and shape not only the nature of the tourist experience but also inscribe a 'superordinate' position to 'Western tourists' so that the 'power bias, the asymmetry, typically skews in the direction of the guest.' According to Costa (1998), it is the underlying positioning of powerlessness of these destinations that exposes the underbelly of representations of fantasy. The deconstruction of the Western imaginary of Caribbean Paradise therefore challenges the perpetuation of the exploitation of these destinations for Western consumption (Sheller 2003:55)

Looking through a postcolonial lens, Echtner and Prasad (2003:679) contend that even decades after most Third World nations have achieved political independence from colonial rule, yet they are still being controlled through colonial discourses evident in tourism marketing that portray these destinations in power relationships that privilege the West. They point out that the term *Orientalism* coined, by Edward Said, describes the signifying practices of Western construction of the East as exotic, backward, and primitive. The term *tropicalisation* more specifically denotes discourses of relations between temperate and tropical zones such as the 'Caribbean and Latin America' that portray these landscapes as 'sensual, and luxuriant and the peoples as pleasure-seeking and profoundly idle' (Echtner and Prasad 2003:667). For Echtner and Prasad, the main context of Third World tourism marketing has been determined by Western mythologies that have mainly contributed to the organic and autonomous images of these countries. They uncover three dominant myths of Third World tourism representations as the Myth of the Unchanged, relating mainly to Eastern cultures, that depicts the timelessness of the Orient, the Myth of the Unrestrained, which is invoked in the representations of sun, sand, and sea tropical beach resorts, and also the Myth of the Uncivilised, used mainly to describe pristine, frontier, and untamed landscapes and regions.

Echtner and Prasad (2003:672) claim that the Myth of the Unrestrained, as used in Caribbean tourist imagery, employs a romantic discourse of 'colonial exploitation,' which suggests that the landscape and the people are totally accessible and willing to serve the fantasies and hedonistic desires of Western tourists in a comfortable, liminal environment. In other words, the Myth of the Unrestrained produces tropical spaces that provide the backdrop for tourists to pursue existential experiences that are not possible in the ordered, civilised, yet more restricted societies of their home countries.

The current marketing slogan of the Caribbean Tourism Organisation, 'Caribbean, everything you want it to be,' apparently reinforces this mythical appeal of the region as malleable, accommodating, and without any limitations to tourist enjoyment. In this context, the Caribbean provides space where the tourist may resolve whatever dissonance occurs in everyday life of developed societies. It strikes an existential note that suggests that the individual tourist has the power 'to be' and to 'become' in imagining and shaping an experience that can only be acted out, upon, and within the tropical space. This has considerable appeal and currency for tourists who may view such existential experience empowering (Arnould and Thompson 2005:675). However, these meanings of the holiday experience also serve the function of 'ego-enhancement,' that is, prescriptive, and do not give much credence to alternative modes or meanings of experiencing Paradise (Echtner and Prasad, 2003). Paradise is therefore not only the attributes of iconic palm-fringed beach and smiling natives, but it is also enacted in the power to ascribe and to attribute value and to also devalue. The ideological power of this positioning is the notion that no other way of seeing or experiencing the region is legitimised.

The prevalence and the evocation of colonial tropes in the Western imaginary of travel and tourism, according to Echtner and Prasad (2003:679), may be linked to cultural strategies to manage and exercise this power of signification in the holiday experience. In the case of the Caribbean, the dominance of the colonial motif arguably may be justified in terms of the realities of a shared historical experience. However, in terms of a postcolonial critique of this discourse, the issue is not merely the invocation of this colonial past but how this invocation works in terms of the representations of these destinations and their holiday experiences. An analysis of the colonial motif as a discursive strategy of representing the holiday experience is illustrated in the extract below from a travel article on Grenada that was published in the *The Observer* in the United Kingdom in 2000.[5]

> Imagine a Britain with sunshine nearly every day—a Britain so warm that palm trees grow on the edge of all beaches. Discounting the effects of global warming, the closest you can probably get to this Britannic idyll is by making the nine-hour air journey to Grenada in the Caribbean. In our English-speaking former colony, the red telephone boxes are exactly the same as the London ones. The Queen still reigns. The policemen have 'ER' badges on their caps, more than matching the British bobby for smartness, right down to the black, shiny boots.

From the first sentence of this extract, the Caribbean is being idealised in terms of the mother country, Britain. The writer invites readers to 'imagine,' that is, to impose a way of seeing the island destination of Grenada as an idealised Britain. In other words, the writer is suggesting that the Britannic idyll is characterised by the imperial signifiers of red telephone boxes, 'bobbies,' and their comparable shiny boots that is complemented by the attributes of the warm weather and tropical landscapes afforded by the destination.

What is important in this representation is how value is ascribed. Indeed, the writer distinguishes warm weather, beaches, and palm trees as the value-added benefits of this island, but these are valorised within the context of their colonial association. The writer is therefore entirely confident in asserting as fact that 'the Queen still reigns,' erasing the reality of the island's political independence. The colonial motif therefore operates as a mode of inscription consistent with the ways early colonisers saw the New World, as a place to inhabit and to stamp their civilisations on, and to claim in the name of the ruling monarch. The imperial *I* or *eye* of the visitor establishes the claim to legitimacy and authority in seeing the world. But in the context of travel, this quest is oriented for a search for an idealised Britain. Although there is the notion of escape and fantasy, noted in the invocation of the imaginary, yet this is not an escape to an unknown world or to the unfamiliar that has merit for and by itself. Instead, the escape is to an extension of Britain endowed with the attributes of tropical weather and landscapes.

This discourse therefore relies on the identification of imperial signifiers to prescribe a way of seeing and experiencing the island. In this respect, the discourse constructs the notion of the holiday ideal as related to the transfers of British culture within the landscape of the tropical. The landscape of Grenada is devoid of local cultural meanings: it is only an attribute to escape to and to enjoy while the context of British culture and civilisation provides the real meaning for the experience. The outcome is that Grenada has no distinct brand identity apart from its colonial heritage. Stripped of these colonial vestiges, it becomes essentially a place endowed with good weather and pleasing landscapes.

The writer appears to be grappling with the realities of the landscape that are not easily conscripted into the imaginary of a tropical colonial British landscape. There is some dissonance in the tourist gaze in reconciling and disciplining the island's landscape into the Western imaginary of tourist experiences.

> But the island is still poor. While no one appears to starve, very few people reach European standards of living. Most British holidaymakers could buy the entire stock of any of the St George market stalls without noticing the outlay. Some holidaymakers will rejoice in the fact that Grenada is 'unspoilt'; others will feel slightly uncomfortable that the locals benefit so little from tourism and trade. There are signs, however, that international investment will start to enrich the local economy and provide more restaurants and other facilities for affluent Western travellers.

The contradiction of the ideal Britain transported to this tropical idyll and overt evidence of poverty apparently poses an epistemological dilemma for the writer. There is an acknowledgment that overt markers of poverty and primitiveness are consumed in the tourist gaze and the consequence of this ideal for the locals momentarily disrupts the Western imaginary of Paradise. The use of the modifier *still* before the descriptor *poor* works at this point in the text to suggest that some expectation was not realised. It proposes a query of the disturbance of an ideal, interrupted by a reality that is not entirely congruous and justifiable. While at the beginning of the article there was the intent to identify the familiar, as well as the similarities between Britain and the island, now there is the depiction of the contrast of living standards.

Interestingly, the poverty of the local people is not presented by this writer as a failure of the civilising mission of the imperial past, but rather it is presented as an endemic issue for which foreign assistance will provide the solution. Since the evidence of poverty is disturbing, the tendency of the discourse is to ease this dissonance in the minds of readers by reducing it to the point that the holiday experience is enabled without much consideration of local poverty. This extract also exposes another discursive strategy that involves the promotion of tourism development as the logical solution to issues of underdevelopment and poverty. In this discourse, tourism is situated here as an industry of exchange where the provision of luxury facilities is linked directly to improvements in the local economy.

The representation of tourism in terms of its exchange value works to legitimate its operation in less-developed countries and attempts to resolve the tensions that may arise between an industry that promotes affluence and well-being in settings that bear the scars of deprivation. But the conceptualisation of tourism as social exchange does not take into account the inequalities of this relationship. For countries whose economies are heavily dependent on tourism earnings, there are apparent dangers of exploitation and an imbalance in the commitment of resources to develop and sustain tourism industries without commensurate indications of poverty reduction and alleviation in local communities.

It is not being suggested that the textual analysis presented here represents the only way of interpreting the above extract. What is being proposed is that this reading of the text may be interpreted as a way in which colonial discourse was used to construct the holiday image of Grenada. It identifies how British culture was foregrounded and presented as the locus of symbolic value for seeing and experiencing the destination. On the other hand, Grenada has been mainly represented in terms of its landscape where the functionality of these attributes was used to define the holiday experience of the island. In this regard, the image of the destination has been confined within the stereotypic frame of the attributes of sun, sand, and sea and does not provide or offer an appreciation of Grenada's symbolic brand image or identity.

CONTESTATIONS FOR THE CARIBBEAN BRAND IMAGE

The work of both Sheller (2003) and Thompson (2006), detailing the historical evolution of the Caribbean imaginary, show that representations of the region have never been static. In their critical examination of historical documents, narratives, and images, they contend that integral to the imperial project in the region was a studied intent to rework and mould the landscape to reflect Western ideals. The introduction to the Caribbean of various flora and fauna from far-flung regions of the globe, the organisation of gardens and tropical vegetation, the photographing of native women with fruit-laden baskets on their heads all enshrine moments of re-creation, 'appropriation and contestation' of the Caribbean landscape. The Caribbean imaginary was therefore forged and systematically erected as part of the civilising mission in the New World. Sheller (2003:37) also points out that there have been both continuities and shifts in the construction of the European imaginary that relate to 'changing modes of economic enterprise and consumer envelopment in the tropics.'

Thompson extends this particular investigation of the tropicalisation and construction of the Caribbean picturesque in an in-depth examination of the photographic images of the Bahamas and Jamaica that preceded the development of mass tourism in the region. She observes that in keeping with popular imaginings of the Caribbean at the time, the main images of the region were those of premodern, natural, untamed, pristine landscapes that were untouched by 'progress' and the passage of time. But this imaging was not without detractors because some 'travellers expressed discomfort with the wild, tangled and uncultivated displays of nature' (Thompson 2006:116). The apparent contestations and struggle for the meaning and visual markers of Paradise clearly complicate the notion of the stability of these representations. From as early as the 1920s, the tension between the need to modernise Caribbean nations in order to provide comfort and satisfaction to tourists and to also maintain notions of premodernity and primitiveness was, according to Thompson (2006:140), normally resolved in the discourse by decentring and masking those civic 'improvements' such as paved roads and electric poles in photographic images and postcards.

As noted in the earlier discussion of the newspaper extract, there is a contrapuntal tension in reconciling the primitive ideal with evidence of lack and poverty. Likewise, this apposition of meaning is also apparent in reconciling destination images of premodern exotica and aspiring, modern, competitive nation-states. The concomitant meanings of Paradise that invoke notions of the timelessness of the region, as somewhat curious and not entirely on the cutting edge of development, are problematic as well for contemporary Caribbean tourism promoters. This was illustrated by the Secretary General of the Caribbean Tourism Organisation (CTO), Vincent Vanderpool-Wallace, in an address he made at an international marketing conference in 2005.

> I remember a friend of mine who might be in this audience telling me some years ago that if the world ever came to its end, he would want to be in the Caribbean. Needless to say, as a Caribbean man, I was most flattered to hear this and thanked him profusely for this obvious admiration of our physical beauty and stimulating culture. He replied: "No, No, No . . . you don't understand. If the world ever came to an end, I would want to be in the Caribbean because everything always happens 10 years later in the Caribbean." *That's the first impression of our brand that we have to fix.* (Emphasis mine)[6]

It seems that the Caribbean promoters are also engaged in the battle to define and construct 'Caribbean Paradise' and may not simply cede the right to ascribe and designate meanings to consumers. In order to maintain competitiveness in the fierce battle for the minds and bodies of tourists, Caribbean marketers may be keenly aware of the need to differentiate and add value to the Caribbean brand in order to gain the edge in the market. In the same address, the CTO secretary general therefore advocates innovation and creativity in the application of information technology as the way forward to gaining the advantage by satisfying and even surpassing customer expectation. In clarifying this approach he states:

'So how will we further differentiate ourselves in the Caribbean? By using information technology to deliver service that makes a difference. I have come to believe that technology should be used to make personal service even more personal, not to replace it.'

The paradox of divergent contestations of the Caribbean as sites of primitiveness, the laid-back and quaint, along with cutting-edge tourism, may not be simply resolved or 'fixed' with technological savvy. Technological innovation is such that it is also reproducible and may not differentiate or put the Caribbean ahead of competitors for an extensive period of time before they 'catch up.' Along with technological modernisation, there is also a need for a corresponding epistemic shift in imagining the region that rationalises new ways of representing holiday experiences offered by the region that serves to undermine substitutability and create value added meaning for the Caribbean brand.

However, the notion of subversion and challenge to hegemonic touristic discourses of the region is also problematic in the search for an alternative ontology that resonates with credence and conviction. In their deconstruction of the Myth of the Unrestrained in the Western construction of tropical paradise, Echtner and Prasad (2003:672) highlight the silencing and the erasure of the realities of the harsh conditions of life in regions where there are 'hurricanes, cockroaches and poisonous snakes.' However, it is arguable whether a strategy that is inherently more 'realistic' and ethical will successfully market the brand. The question is also the extent to which spotlighting local realities or demystifying paradisiacal representations is tantamount to disrupting the 'tenacity of colonial discourse in tourism marketing' (ibid., 679). It is dubious whether such counterdiscourses could be reasonably effective in redressing the reductionism of touristic representations that 'demean Third World places and people' and also enhance the region's competitiveness. Echtner and Prasad (2003:680) therefore summarily conclude that they were unable to articulate modalities and agencies for subaltern voices to speak in order to redress these imbalances. Shepherd (2007:206) also acknowledges the challenges in redressing normative epistemologies of postcolonial Caribbean societies: 'The question of how to construct a Caribbean identity that will negate the images and representations of the colonial and present eras and challenge the epistemic violence of the imperial project remains an important and pressing one.'

She argues that there are possible 'emancipation strategies' available in the recovery and dissemination of positive images of survival and achievement that could contribute to a 're-imaging' of Caribbean peoples.' Shepherd (2007:207) contends that the development of more revisionist projects, such as have been achieved by Caribbean scholars in the disciplines of history and political economy, 'are required in order to provide the ammunition for a 'counter discourse to [the] Eurocentric texts . . .'

But counterdiscourses or 'reverse-discourse as an oppositional practice,' as noted by Parry (1994:172), risk reinforcing the essentialisms of assumptions of master discourses of the coloniser/colonised. The privileging of native voices to the total exclusion of others may in turn mimic or replicate the modality of knowledge systems under which subalterns have been repressed. The creation of countermyths that only reify the subaltern may also reduce the complexities, polyvocality, and hybridities of Caribbean culture and societies. The term *creolisation* as a critical, conceptual motif of the dynamic evolutions and history of the region denotes the 'interrelations and, indeterminacy and cross-cultural hybridity of Caribbean identity and cultural expression.' It also represents endemic cultural practices of resistance that have emerged from the 'crucible of Caribbean slave and post emancipation societies' in the quest for subaltern survival (Allen 2002:47). Although creolisation practices are antithetical to the imperialist view of the Caribbean, and represent a 'significant ideological shift in the decolonisation of the region,' it is not simply a process of displacement or an insular retreat into myths of nativisation and indigenisation (Bolland 2002:18). According to Allen (2002:50), the term *creolisation* is yet 'fraught with ambiguities,' but these meanings have coalesced so that 'etymologically it now denotes fragmentation, obscurity, possible invention or corruption and adaptation.' Allen enumerates core genetic characteristics of creolisation as a form of subversion and resistance that acknowledges the importance of cross-cultural encounters, *negotiating between identities and forces,* rejection, adaptation, accommodation, imitation and invention' (emphasis mine). She explains that 'creolisation also values nativisation or indigenisation, but this is not fixed but is part of a dynamic process of change rather than an alternative or reverse 'Othering' process.' As such it is not simply the articulation of a reverse discourse. Caribbean creolisation discourse therefore complicates binary differences in the construction of the 'Other' by providing strategies for varying modalities of resistance that is simultaneously ambiguous and syncretic.

The apparent dominance of the Western hegemonic discourse in the imagery of Caribbean destinations suggests that the creolisation process may not have penetrated this area of Caribbean economic, social, and cultural life. The prevalence of Western tropes in the self-representation of Caribbean countries in promotional and advertising material in brochures and the official tourism Web sites of regional tourist boards may presuppose the charge that tourism promoters have been 'passive actors who are helpless to do anything but reproduce the structures of their own subordination'(Parry, 1994:176). But Thompson (2006:281) contends that there have been instances of local contestations that evidence attempts to counter 'prevailing touristic constructions.' She highlights the example of the 1970s advertising campaign of the Jamaica Tourist Board (JTB), under the premiership of Michael Manley, that declared to the world: "Jamaica: It's more than a beach. It's a country."

Thompson claims that this was an assertion by the Jamaica Tourist Board at the time to assert 'geographical specificity in the face of the representational tyranny of the newest tropical icon—the beach.' However, this assertion of the Jamaican space and identity may also be read as the nativi-sation or indigenisation stage of the process of creolisation that involves the recovery of the subaltern voice. Interestingly, the syncretic or adapta-tion strategies of creolisation are observed in the subsequent 'One Love' advertisements in the 1990s that adopted the reggae music and melody but revised the original lyrics of the widely popular 'One Love' song by Bob Marley. Cooper (2007:229) establishes the clear link and antecedents between the 1970s campaign and the later campaign of the 90s:

> 'Indeed, the message of the revolutionary 1970s advertisement, "We're more than a beach, we're a country," seems to have penetrated the mar-keting strategies of the 1990s with respect to reggae.'

However, Cooper (2007:223) also exposes the 'adulteration' of Bob Marley's lyrics that desensitises the meaning of the original text that was an exhortation for local unity, to accommodate the 'sun and fun priorities of the marketing experts.' The revised version of Marley's lyrics used by the Jamaica Tourist Board (JTB) Cooper evaluates as unquestionably positive as it encourages visitors to

> Feel Jamaica move you
> Warm Jamaica touch you . . .
> Come to Jamaica and feel alright[7]

In spite of the toning down of the note of protest in the revision, Cooper (2007:223) also supports this adaptation in achieving global cultural affir-mation and appreciation of the 'remarkable creativity' and cultural identity of the Jamaican people. While the revision of the lyrics by the JTB maintains and promotes a conventional sensuous experience of Jamaica, it seems that an attempt was being made to syncretise conventional notions of the Jamai-can holiday experience by opening up the unique cultural landscape of the country through the medium of reggae music.

The extract below, taken from a review of the book *Bass Culture,* which chronicles the historical evolution of reggae music, illustrates the ascen-dancy of culture in engaging and opening up new cultural pathways and experiences.

> Culturally speaking, reggae and its accompanying musical baggage are Jamaica's great gift to the world. Among the island peoples of the West In-dies, only the Cubans and the Trinidadians have come close to the export achievements notched up by the pop musicians and producers of post-in-dependence Jamaica. *Nobody outside America has done more to influence not just what we listen to, but the way we listen to it.* (Emphasis mine)[8]

The writer readily expresses his own views on the music when he states that reggae music is 'a gift to the world.' He sets out to build on this main premise by highlighting the contribution that the music has made to other musical forms as well as to popular culture. Here the writer strikes a celebratory tone in his representation of the origins of reggae music. He attributes the music with powerful influences, triumphing over the insignificance of its roots, so that the music is portrayed not just as another musical form but as a discourse of achievement and overcoming adversity.

Based on this representation of reggae music as emerging from outside the mainstream, it is endowed with a subversive appeal that also accounts for its potency and attraction in Western popular culture. However, this celebratory position also resonates with Bhabba's notion of hybridity, which suggests the undermining of fixed polarities and differences between cultures (Sheller 2003:193). This influence is particularly demonstrated in the area of popular culture that affirms and encourages the adoption of modes of cultural resistance. The description of the origins of reggae and the powerful responses it has evoked attests to an indigenous art form that is undeniably authentic. This is an attribute of the music that is valorised and lauded by the writer. In this respect, reggae music is presented as 'cultural capital,' adding value and enrichment to Western cultural consumption. This discourse of popular culture therefore acknowledges merit to the artistic fringe and celebrates, accepts, and also appropriates reggae music.

These positive cultural attributions are also rich in symbolic meanings that may be beneficial for Caribbean destinations. In some respects, the benefits of these positive images may appear to be appealing mainly to those segments of the market that are interested in culture. However, even beyond its segmental appeal, the valorisation of the Caribbean's cultural indigenous expressions and cultural exports may be harnessed to build brand identity based on an attributed symbolic value of the Caribbean cultural experience. Symbolic value of the region is achieved when the image of the destination is represented as more than a collection of physical tourist product attributes but is also culturally potent and dynamic.

The appropriation of cultural resources is particularly compelling in creating new and differentiated cultural conceptions of holiday making that disrupts and undermines closed, cognitive frameworks of seeing and experiencing holiday destinations. By drawing on the cultural power of the Caribbean to construct symbolic and new meanings, the mere functionality of the sun, sand, and sea stereotype that has defined and departicularised Caribbean destinations may be negotiated to build distinct, symbolically rich brand images. Cooper (2007:233) strikes a note of triumphalism for the differentiating power and sustainability of Jamaica's cultural resources when she declares: 'This is the cultural heritage that Jamaicans can afford to share. It is a renewable resource that cannot be totally consumed, not even by the most voracious tourist.'

It is the appeal and the potency of the creative cultural resource that contributes to symbolic value and brand identity for destinations and engages in a dynamic process of destabilisation of hegemonic discourses of the Caribbean Paradise. This may be related to the quest to resist totalising and normalising master discourses that dominate representations of the Caribbean. Relatedly, creolisation by promoting syncretism and hybridity may facilitate differentiation that serves the purposes of building distinctive, value-added brands. Although using the term *glocal* rather than creolisation to describe the attempt to modify and disrupt conventional images, Therkelsen (2003:145) reasons that the recognisable, conventional, and global image of the destination can be used as conduit for the introduction of new meanings from the local cultural context that may serve the interests of brand differentiation:

> Utilising well-known image components can, furthermore be seen as way of supplying the tourism destination with additional meanings and value, as well as mediating something possibly unknown to the target audience (holidaying at a given destination) by placing it within a well-known frame of reference (the meaning associated with the place in general). Eventually this process may help establish the product firmer and more nuanced in the mind of the potential tourist.

In other words, leveraging for the brand may be realised through a strategy that 'uses the cliché as a hook on which to hang more detail, so that this clichéd identity may be reshaped and given greater complexity' (Therkelsen 2003:144). This line of argument provides the economic rationale behind creolisation strategies in imaging and marketing the Caribbean as it clarifies the benefits in offering differentiated, value-added, symbolic meanings for the region's brand image and identity. According to Blain, Levy, and Brent Richie (2005:331), since many destinations offer similar attributes, effective destination branding necessitates a unique selling proposition that competitors may attempt to copy but are unlikely to 'surpass or usurp.' They conclude that differentiation strategies in branding are likely to contribute benefits to the destination by reducing substitutability, as well as 'limiting discounting' and 'preventing slippage into the maturation phase of the destination lifecycle' (Blain, Levy, and Brent Richie 2005:331).

STUDY METHOD

While this discussion of Jamaica's destination campaigns provided examples of varying degrees of creolisation strategies, an examination of the home pages of the official tourism Web sites of the countries of the Anglophone Caribbean will also aid this enquiry in determining the extent to which there may be such manifestation in current branding activities in the region. Parry (1994:173) suggests that 'traces, inscriptions and signs of resistance' may be identified in narratives and practices that indicate defiance and identity assertion. So for the purposes of this enquiry, creolisation has been operationalised as differentiation strategies that are evidenced in the subversion of traditional stereotypes and the attempt to foreground national identity and culture. Such strategies also include an attempt to revalorise and subvert so that the power to ascribe meaning is located within the destination rather than the visitor. In achieving this goal, this enquiry will first examine the ways in which Anglo Caribbean destinations are differentiated based on traditional image constructs of the region as sites of Paradise. The second level of analysis addresses the examination of less dominant, underlying 'traces' of opposition, subversion, and adaptation that may be evidenced in the narratives and images of the promotional material presented by the national tourism organisations.

The destination Web sites used in this enquiry were accessed through the official Web site of the Caribbean Tourism Organisation (CTO) located at http://www.caribbean.co.uk/info/index.html.en-GB. From this location specific links are provided to the designated official Web site home page of the 18 Anglo Caribbean destinations that are examined in this study. With the growing popularity and pervasiveness of the Internet, the promotional material on the official tourism Web sites of national tourism organisations has become a focus of academic research of the branding and image-management practices of tourism destinations (Choi, Xinran, and Morrison 2007; Lee et al. 2006). Home pages may be particularly useful as units for content analysis because they highlight the most salient messages, images, and concepts that have been selected by destination managers to represent the image and brand of the destination. This process of selection is also referred to as framing, which is aimed to influence meaning making through the way in which the information is organised and presented. Santos(2004:124) suggests mediated tourism messages provide dominant frames for tourists that form the basis for 'the representational dynamics used within tourism discourse.' As such, home pages represent a configuration and construction of meaning that provides a dominant frame of reference. However, there are also some limitations in the use of home pages as units of analysis, mainly that these sites tend to be updated regularly. While this is effective and handy for tourism marketers in getting current information to the market quickly and in 'real time,' it also introduces inconsistencies in the process of analysis over a period of time. Nevertheless, core values of attribute selection and messages tend to remain stable, and so some level of continuity can be maintained in the research process. Some home pages, such as those of Barbados and Trinidad and Tobago, are designed mainly as gateways to more information presented in more detail provided on other pages of the Web site so that there are varying levels of information presented across the various home pages. Another limitation is that this analysis and discussion do not infer to the actual effect of these framing strategies in terms of consumer reception but only expose the ways in which meaning is constructed and organised in the texts and images.

STUDY RESULTS

It may appear a daunting task to identify unique features to differentiate the many destinations within the Anglo Caribbean region. To date, their similarities as sun, sand, and sea destinations have been the basis for their renown in the travel market. These images have been successfully exploited but in turn have also created the impression of the Caribbean as offering the same basic experience differentiated only by a change of geographical location or country name. Consequently, representations across destinations have been mostly flat and homogenous in their depiction of similar landscapes, types of people, and activities.

The dilemma this poses for individual destinations is the task of creating a brand identity that shows the destination as offering 'something extra' in comparison with competing destinations. Often competitive advantage is achieved in terms of which destinations have a 'deep' marketing budget to create high-profile visibility and awareness of the country in the marketplace rather than being able to benefit from a distinct, valued brand identity. This may mean that a high advertising spend has to be maintained in order to keep awareness of the particular destination at top of mind.

Although there is a general blurring of differences between Caribbean destinations in the narrative and images of the various home pages, it is possible to pinpoint their relative brand identities, that is, the main holiday experiences and image constructs that best define them. It is true that whatever label may be claimed for one can also be claimed for another to some degree, but it may also be argued that each destination can be classified in terms of key attributes. Table 1.1 following presents a qualitative typology of the various destinations of differentiated image constructs based on the content of the various home pages.

Four main dominant image constructs or brand descriptors have been identified to distinguish destinations. These are Nature/Eco, Paradise/Recreational, Elite/Exclusive, and Cultural Destinations. However, it is important to remember that all eighteen countries also include aspects of these various constructs and that no one construct is entirely exclusive to a destination.

The image constructs or attributes of the destinations itemised indicate the areas of distinctive brand appeals that they portray on their home pages. However, the promotion of a destination brand image also involves capturing the essence of the destination's experience and its effects on the consumer. In finding the unique trait that is useful in building the brand image, the aim is to offer to the consumer a bonus that is based on a 'recognisable differentiated value.'

Table 2.1 Main Image Constructs Differentiating Caribbean Destinations

Image Constructs	*Destinations*	*Key Characteristics*
Nature/Eco Destinations	Belize, Dominica, Guyana, Montserrat	Mainly represented as underdeveloped with attributes such as rainforests, fragile biodiversity and generous endowments of flora and fauna
Paradise Destinations	Antigua & Barbuda, Bahamas, Barbados, Grenada, St. Lucia, St. Kitts & Nevis, Tobago, Anguilla, Bermuda, Cayman Islands	Offers the typical recreational sun, sand and sea holidays. Featuring deluxe hotels, modern facilities and many varied holiday activities.
Elitist /Exclusive Destinations	Mustique Anguilla, St Vincent & the Grenadines, Turks & Caicos, Cayman Islands, British Virgin Islands	Emphasises that the remoteness/retreat or lack of tourists on the destination. Focuses on seclusion, and hidden locations with not many tourists, not for the mass tourist market
Cultural Destinations	Jamaica, Trinidad & Tobago	Emphasises indigenous cultural expressions of music, festivals, art, entertainment

Source: Author

This may be particularly crucial for destinations in terms of a marketing positioning because they can maximise the perception that they have something more to offer their customers. Chen and Uysal (2002:987) maintain that competitive marketing positioning is vital for the long-term success of the destination, and this is usually achieved by being able to successfully market a distinguishable physical or symbolic feature of the destination. This is particularly crucial for Caribbean destinations that have few clear differences to distinguish them based on their physical and functional attributes.

The second level of this examination of the home pages focuses on the identification of less-dominant traces of resistance and oppositional strategies of the narratives and images of the promotional material. This reading of the texts examines the rhetorical strategies of slogans, choices of images as well as their positioning in terms of their foregroundings, decentring, and juxtapositions. It also considers the narration of histories and accounts that presents indigenous interpretations of Caribbean history, identity, and culture (Shepherd 2007:19). The aim is not to give an exhaustive and detailed analysis and account of all the home pages but to highlight some examples to illustrate and substantiate the theoretical suppositions of this discussion.

Generally, this investigation did not identify overt, distinct challenges to conventional images on the home pages of the destinations under examination. However, there were some instances and examples of emergent oppositional differentiation strategies that were presented to merit discussion. These strategies may be termed as approaches that demonstrate a counterdiscourse that seeks to accentuate the local and to indigenise the holiday experience by foregrounding the cultural landscape of the destination. Is also involves the adaptation or syncretisation of existing conventional images by shifting their usual meanings to introduce new ways of seeing or experiencing the destination. In other words, a syncretic/adaptation strategy uses the clichéd meaning to set the context to extend the established view of the destination. It therefore destabilises conventional meanings by mixing and confusing the stereotypes with new concepts and associations. For example, this approach is seen in Antigua and Barbuda's slogan 'the beach is just the beginning,' which suggests that island's iconic image of having a beach for every day of the year is being somewhat reconfigured. The implication of this slogan is that Antiguan tourism offers new ways of appreciating the island beyond the iconic beach.

Significantly, many of the destinations' home pages also depict locals in tourist settings enjoying traditional activities normally engaged in by tourists. For example, one of the images on Bermuda's home page shows what could be assumed to be a young 'native' boy running on the beach. The British Virgin Islands (BVI) also depicts locals enjoying water sports as well as the traditional image of White, Western tourists on the beach. St. Kitts and Nevis highlights a picture of a young man captioned by the words *I am Kittitian,* which serves to assert his identity as integral to the destination brand. The inclusion of locals and other ethnic groups in traditional tourist settings serves to undermine the polarity and binary designations of the 'Other.' This is reinforced by the noticeable absences or decentring of images of 'locals' positioned in conventional service encounters with White Western guests across the majority of the home pages of the destinations.

Other examples of differentiation/oppositional strategies introduce a subtle reverse discourse by encoding and introducing local cultural meanings and iconic national symbols in the presentation of the destination's holiday experience. Several destinations therefore have chosen to incorporate national symbols for their tourism logos. Belize includes the national bird, the keel-billed toucan, alongside the destination name and the slogan 'Mother Nature's Best Kept Secret.' Grenada also includes their iconic referent of the nutmeg but also uses this to establish specified holiday experiences for the destination by creating tags such as 'Spice Holidays,' 'Spicitivities,' and 'Spice Up Your Life in Grenada, the Spice of the Caribbean.' Jamaica also employs the technique of encoding cultural cues by foregrounding of the image of a colourfully decorated urn captioned as 'Wassi Art Pottery' positioned on the masthead of the official destination home page. The likely unfamiliarity of this local artistic craft may have been staged here to introduce new and nuanced meanings into the conventional holiday images of Jamaica and also serves to encourage the valorisation of indigenous artistic, cultural expressions within the overall holiday experience on the island.

In differentiating the island from the other destinations in the region, the narrative promoting Dominica invites tourists to 'visit our island and defy popular notions of a Caribbean vacation.' With this statement, Dominica sets itself apart from the more conventional sun, sand, and sea tourism but also pushes the boundaries of the definition of Caribbean holidays to include nature/eco-holidays. More significantly, Dominica's home page also features a hand-painted portrait logo of twinned mountains peaks on which a feminised, native facial feature is etched and superimposed. This could be read discursively as an attempt to disrupt conventional representations of scenic, picturesque landscapes so as to visually present alternative meanings of the holiday experience on the island by foregrounding the indigenous features as a signifier of an authentic cultural landscape.

However, an important caveat should be noted that consensus on the appropriation of differentiation strategies may not be easily achieved among stakeholders at the destination level or regional level. Cooper (2007:230) recounts the internal disputes that occurred in Jamaica when the 'One Love' advertisements were first proposed, noting the angst of some local commentators on the introduction of images of Rastafarians and use of reggae music in promoting the destination. This author was employed at the Jamaica Tourist Board at the time of the launch of these advertisements and recalls that the responses of stakeholders were usually divided between antipathy and enthusiasm. Clearly, efforts to brand Caribbean destinations will depend to great extent on the ability of tourism destination managers to galvanise local acceptance and support for such promotional strategies.

Caribbean scholars such as Rex Nettleford (1995:84) have long advocated the appropriation of the Caribbean's cultural expressions as a modality of development, negotiation, and liberation as the region interacts with the world. He posits that

> Music, dance, religious expression, language, literature, appropriate designs for social living are the structural products of the Caribbean's creative impulse. They serve as guarantees of inner space meeting outer space in a dynamic existence that turns on creative rather than disintegrative tension.

Table 2.2 Description of Differentiation/Oppositional Strategies

Differentiation/Oppositional Strategies —Adaptation/Syncretic and Nativisation/Indigenisation	*Destination Examples*
Undermines the stereotype by high-lighting its limitations as a fulfilling holiday experience. Shifts the locus of the vacation experience from the ocular, functional attributes of sun, sand and sea to the cultural land-scape of the destination	Antigua's slogan—'the beach is just the beginning Bermuda'—Image of young, local boy running from the beach into the ocean Bahamas—Picture snapshot of an apparent local wedding in tourist set-ting, non white surfers BVI—images of locals on the beach St. Lucia—images of local festivals and carnivals Dominica's slogan and logo that suggests a 'nativised' concept of the holiday experience with ethnic features superimposed on twinned mountain landscape Belize's use of national symbol of keel billed toucon bird in tourism logo Grenada's use nutmeg in tourism logo Jamaica's use of 'Wassi Art' pot-tery superimposed on the header of Homepage

Source: Author

> This phenomenon is best described by the phrase inward stretch, outward reach, neither of which activity is possible without a sense of space.

In this polemic, Nettleford employs the aphoristic phrase 'inward stretch and outward reach' to define a dialectical stance of an ongoing self-discovery and knowledge creation that interacts with while creatively undermining opposing knowledge systems. Taken within the context of the project of Caribbean tourism development, particularly image management, this dialectic offers some guidance in negotiating the tensions between the Western imaginary of the region as well as local aspirations towards nationhood and self-determination. 'Inward stretch, outward reach' encourages the creative

energies, confidence, and liberation within the region to innovate and construct new knowledge systems. It is a dynamic process of creation that is not
insular in purpose or in practice but also seeks out avenues for influence and
engagement in the wider world. Taken within the context of this discussion
of touristic representations of the Caribbean, 'inward stretch, outward reach'
suggests that the region should continue to engage reflexively in interrogating
the normalising and totalising discourses that that often reduce the particularity, identity, and voice of Caribbean peoples. Moreover, it also emphasises
the power of creative subversions of meaning in order to perpetuate epistemic
shifts that contribute to and enhance the brand image of the region.

To constantly assert that prevailing stereotypical images of Caribbean destinations must continue to be produced and reproduced because it is what
the 'market wants and needs' is to fail to recognise that such market wants
and needs are constantly being revalorised, repackaged, and reinvented.
Increasingly, studies on consumer behaviour are contending that branding is
a cultural process that is performed in terms of cultural codes and ideological discourses (Schroeder 2006:3). Hannerz (1996:156) argues that within
market frames there are dynamic movements of values, tastes, and skills that
shape consumers tastes. He contends further that the 'market as a frame for
the flow of culture operates' as a 'way of socialising people into consumption
patterns.' This presents manifold opportunities and occasions for Caribbean
cultural influence and negotiation. If there is a marketing imperative for the
region's tourism promoters, it may well be that, along with commitments
to technological development for the consistent and unwavering delivery of
tourist service quality and satisfaction, there should also be a mandate to
interrogate the cultural context of consumer consumption so as to more skil-
fully leverage and differentiate the Caribbean tourism brand.

Ongoing examination of the promotional material of Caribbean destinations is therefore advocated in order to expand and sharpen insights on
differentiation strategies that are being utilised or could be employed by
regional tourism marketers in building meaningful, symbolic brand images
for their destinations. Research should also focus on consumer reception
to test the extent to which such meanings are being decoded, accepted, or
resisted by main target markets. The benefit of such a research agenda will
be to enlighten and perhaps encourage regional marketers to take bolder
steps in utilising potent cultural resources in distinguishing and re-visioning
the Caribbean tourism brand.

CONCLUSION

This chapter highlights the dominant invocation of colonial tropes in mediated touristic representations of the Caribbean. It is argued that colonial
tropes and practices of 'Othering' remain fundamental to the meanings and
rationale of seeing the Caribbean and the travel experience in the region. This

study proposes that stereotypical representations of Caribbean destinations may be undermining competitiveness and reducing the chances of enlarging the scope and range of holiday experiences and destination appeal in the marketplace. Of course it suits prevailing travel interests to maintain this ontology, as it provides stereotypical credence and continues to generate and stimulate a steady demand for travel characterised by low involvement behaviour in the marketplace. It is therefore difficult for destinations to challenge or contest the economic basis for representations that convincingly typify 'what the consumer wants and needs.' This economic logic provides the ground to legitimise and naturalise the current representations that are dominantly purveyed in the media and also in the self-representation of promotional and marketing communication messages of the Caribbean brand.

In critiquing the limitations of existing sun, sand, and sea images of Caribbean destinations, this study has explored strategies that may be further explored to enhance the regional brand and continue to sustain the tourism industry in the region. The discussions and findings of this chapter indicate that the domain of culture and local folkways open up avenues for the revaluation and retextualisation of Caribbean destinations and the holiday experience. It is within this sphere of influence of creolisation and indigenisation that alternative epistemologies of the region may be engendered and dispersed (Sheller 2003:190).

In a postcolonial critique on the political and cultural identity of the Caribbean, Lal (2000:236) articulates this vision of subaltern emancipation in his construction and reinvention of the region that incorporates the realities of colonial history but is not circumscribed and limited by these antecedents.

> Thinking of the Caribbean as a civilisation offers a number of possibilities for an emancipatory politics that may otherwise be barred, and allows us to envision a unique space . . . a Caribbean civilisation suggests a way of contending with a legacy of colonialism, suffering, and oppressive knowledge systems that leaves the victims of history not disempowered, as is presently the case with most formerly colonised peoples, but as possessed of the capacity to shape a future that is not simply a version of the contemporary West.

In this rendition of the historical and cultural identity of the region, the normative codes of difference are transgressed. This imaginary offers the invitation to a rediscovery of the Caribbean that is both empowering and liberating for locals and offers numerous possibilities for fulfilling holiday experiences and encounters.

NOTES

1. Data taken from the 2004 Annual Tourism Statistical Report, Caribbean Tourism Organisation, St. Michael, Barbados.

2. Caribbean tourism destinations, hotels, and services have been recipients of numerous awards for excellence awarded by many international and regional groups that are too numerous to list here. In 2006, the Caribbean won eight awards at the World Travel Awards, which have been described as the 'Oscars' of the travel industry by the *Wall Street Journal*. The list of top awards to the Caribbean include World's Leading Island Destination—Turks and Caicos Islands; World's Leading Honeymoon Destination—St. Lucia; World's Leading Cruise Destination—Jamaica; World's Leading Eco-tourism Destination—Tobago Main Ridge Forest; World's Leading All Inclusive Company—Sandals Resorts International, among others. Many of these awards are often highlighted on official tourism Web sites nations as testimonials in a bid to enhance the image and reputation of regional destinations. See World Travel Award website at http://www.worldtravelawards.com/winners2006-1 for list of awards.
3. David Lambert in his review also argued that Sheller's (2003) work, *Consuming the Caribbean*, was significant as a 'vibrant critique that seeks to encourage more thoughtful and ethical relationships with the region.'
4. In May 2005, stories of a missing teenager from the United States who was vacationing on Aruba generated negative publicity for the island. According to reports on FOXNews.com, the teenager's mother met with Aruban tourism officials, who expressed concern that negative publicity surrounding the case could hurt the island's tourism-dependent economy.
5. Extract taken from the article 'Escape: Caribbean Special: Grenada: Spice and Easy' by Neasa MacErlean, published in *The Observer* in the United Kingdom on September 17, 2000.
6. This quote is taken from an address made by the Secretary General of the Caribbean Tourism Organisation Vincent Vanderpool-Wallace in address entitled 'A Focus on the Future: The Challenge of Branding the Caribbean' at the ATME 2005 Conference in Las Vegas, Nevada.
7. Lyrics courtesy of the Jamaica Tourist Board.
8. Extract taken from a book review 'Getting Back to their Roots' written by Lloyd Bradley. The article was published in the *Sunday Times* on August 13, 2000.

REFERENCES

Allen, C. 2002. Creole: A Problem of Definition. In *Creole: Creolisation discourses in Caribbean culture*. V. Shepherd and G. Richards, eds. Kingston: Ian Randle Publishers.
Arnould, E. J., and C. J. Thompson. 2005. Consumer culture theory (CCT): Twenty years of research. *Journal of Consumer Research* 31 (2):868–882.
Blain, C., S. E. Levy, and J. R. Brent Richie. 2005. Destination branding: Insights and practices from destination management organisations. *Journal of Travel Research* 43 (1):328–338.
Bolland, O. N. 2002. Creolisation and Creole societies. In *Creolisation discourses in Caribbean culture*, V. Shepherd and G. Richard, eds. Kingston: Ian Randle Publishers.
Chen, J., and M. Uysal. 2002. Marketing positioning analysis: A hybrid approach. *Annals of Tourism Research* 29 (4):987–1003.
Choi, S., Y. L. Xinran, and A. M. Morrison. 2007. Destination image representation on the Web: Content analysis of Macau travel related websites. *Tourism Management* 28 (1):118–29.
Costa, J. A. 1998. Paradisal discourse: A critical analysis of marketing and consuming Hawaii. *Consumption, Markets and Culture* 1 (4):303–423.

Echtner, C. M., and P. Prasad. 2003. The context of Third World marketing. *Annals of Tourism Research* 30 (3):660–82.

Edmondson, B., ed. 1999. *Caribbean romances: The politics of regional representation.* Charlottesville and London: University Press of Virginia.

Featherstone, S. 2005. *Postcolonial cultures.* Edinburgh: Edinburgh University Press.

Hannerz, U. 1996. *Transnational connections, culture, people, places.* London and New York: Routledge.

Lal, V. 2000. Unanchoring islands: An introduction to the special issue on 'Islands, Waterways, Floways, Folkways.' *Emergences: Journal for the Study of Media and Composite Cultures* 10 (2):229–40.

Lambert, D. 2004. Book Review: Consuming the Caribbean: From Arawaks to zombies. *Progress in Human Geography* 28 (1):138–39.

Lee, G., L. Cai, A. O'Leary, and T. Joseph. 2006. WWW.Branding.States.US: An analysis of brand-building elements in the U.S. state tourism websites. *Tourism Management* 27 (2):815–28.

Loomba, A. 1998. *Colonialism/postcolonialism.* London and New York: Routledge.

Moisander, J., and A. Valtonen. 2006. *Qualitative marketing research.* London: Sage.

Nettleford, R. 1995. *Inward stretch outward reach: A voice from the Caribbean.* New York: Caribbean Diaspora Press Inc.

Palmer, C. 1994. Tourism and colonialism: The experience of the Bahamas. *Annals of Tourism Research* 21 (4):791–811.

Parry, B. 1994. Resistance theory/theorising resistance or two cheers for nativism. In *Colonial discourse/postcolonial theory,* F. Barker, P. Hulme, and M. Iversen, eds. New York: Manchester University Press.

Puttullo, P. 2005. *Last resorts: The cost of tourism in the Caribbean.* London: Latin America Bureau.

Santos, C. A. 2004. Framing Portugal: Representational dynamics. *Annals of Tourism Research* 31 (1):122–38.

Schroeder, J., and M. Salzer-Morling, eds. 2006. *Brand culture.* London and New York: Routledge.

Sheller, M. 2003. *Consuming the Caribbean: From Arawaks to zombies.* London and New York: Routledge.

Shepherd, V. 2007. *I want to disturb my neighbour.* Kingston: Ian Randle Publishers.

Therkelsen, A. 2003. Imaging places: Image formation of tourists and its consequences for destination promotion. *Scandinavian Journal of Hospitality and Tourism* 3 (2):134–50.

Thompson, K. A. 2006. An eye for the tropics: Tourism, photography, and framing the Caribbean picturesque. Durham, NC, and London: Duke University Press.

3 Bob Marley, Rastafari, and the Jamaican Tourism Product[1]

Jalani Niaah and Sonjah Stanley Niaah

INTRODUCTION

At the University of the West Indies' 1997 presentation of graduates, Christopher Blackwell was awarded the honorary degree of doctor of laws in a ceremony in which Jimmy Cliff was also awarded an honorary doctorate (of letters). In a sense these honours are a type of validation from the academy about the contribution of these men's life work and that these contributions are worthy of the highest academic honours. Among the recipients of this honour are Her Royal Highness Princess Alice, His Imperial Majesty Emperor Haile Selassie I, world renowned folklorist the Hon. Louise Bennett-Coverley, and the former West Indian cricket captain Clive Lloyd. Bob Marley has not been the recipient of this type of honour.

Caribbean social anthropologist Barry Chevannes holds that the validation received from one's peers remains high on the scale of personal accolades he received throughout an academic career spanning more than forty years. This academic validation can be equated with a type of tourism product of ideas and intellectuals, where scholars (artists/workers) like destinations vie for attention, feedback, and ultimately consumption by peer reviewers and/or visitors, and high visitor satisfaction constitutes approval and recommendations. The makers of music are a curious type of tourism product largely because of the portability of the medium and its purveyors. Bob Marley therefore now represents much more than a music maker but also, in death, a site of institutionalized memory through the various monuments and tribute sites connected to his memory. Bob Marley has to this extent become one of the most validated cultural icons in Jamaica's history, and by extension this provides a context for assessing his tourist value to Jamaica, especially in the context of his alternative and iconoclastic ideology.[2] Dawes ([1999] 2004) describes the essence of this argument in what he calls a 'reggae aesthetic,' which is embodied in Marley and his music. Since it is well established that the Rastafari faith is a part of the larger presentation of Marley and his ideas, it is not unreasonable to assume that Dawes's 'reggae aesthetic' could also be described as 'Rastafari aesthetic' as well. Bob Marley, irrespective of which aesthetic is chosen, is still viewed as an icon of both reggae and Rastafari. To restate that argument more compellingly, the Rastafari artist Bob Marley is the undisputed king of Reggae. Bob Marley is therefore approved or valid for representations of Rastafari, reggae, Jamaica, resistance, and Africa, among other discourses to which his name has been applied.

This chapter makes no claim about the logic or validity of Marley's iconic and overarching image within modern Jamaican consciousness. Instead, the chapter argues that Jamaican national identity and international appeal are contingent and dependent on the presence and image of Rastafari. Bob Marley has been central in this validation, and since his death his legatees have been able to develop a significant tourist market around his image and legacy. The chapter thus examines the development of the Marley tourism product, especially the Bob Marley Museum as a site and its significance to those who visit Jamaica.

THE PROMULGATION OF A REGGAE AESTHETIC IN TOURISM

Jamaica has had a distinct past in British colonialism and the enslavement of Africans brought to the island as captives. From this has emerged a clear and outstanding international reputation for articulating resistance to a history of oppression and servitude. Within the twentieth century the Rastafari movement,[3] an Ethiopianist-centred pan-African organization established in Jamaica in the early 1930s, largely by Leonard Howell,[4] has emerged as one of the key signals of this resistance. Leading Rastafari elders such as Mortimo Planno were able to guide and instruct young musicians toward singing the teachings of Rastafari, which gave the movement an international voice through the medium of reggae music, arguably the first 'world music' (Connell and Stanley-Niaah, forthcoming). Reggae music is often complexly defined; however, for the purpose of this discussion we refer to the body of spiritual yet political music that emerged out of Jamaica in its post-1962 independence. This music is decidedly different from the more calypso-oriented sounds developed from *mento* and *ska* music, being slower in its pace and having more pronounced drum and bass. The influence of *burru* and *nyabinghi*[5] drummers provided an atmosphere for cultural exchange from which reggae emerged on a continuum from ska to *rocksteady* and then reggae, somewhere between 1966 and 1968. This cultural exchange, recognized by researchers such as Chevannes (1981, 1998), was developing through Rastafari alternative leadership that extended across Kingston, from Wareika Hills in the east, to Trench Town and beyond in the west. It is from this environment that artistes such as Bob Marley emerged to become one of the biggest names on the planet and his Rastafari identity a global symbol of resistance but simultaneously the human hope of 'one love' (Niaah 2006). Marley's face has given to Jamaica a distinct and unique identity that imposes some interesting ironies of history with the face of resistance being the key invitation for visitors to come to take advantage of Jamaican hospitality. This is not dissimilar from the use of slave ports/forts and dungeons in West Africa or the sites of apartheid in South Africa as tourist attractions.

In a tourist promotional documentary aired on the Travel Channel (USA) in 2005 entitled 'Jamaica the Ultimate Tour,' the then prime minister of Jamaica, P. J. Patterson, guided a five-day tour of the island's range of attractions for guests while sending the message 'you will always have a home in Jamaica.' The film opens with a sequence of visual enticements for such would-be guests and/or investors, introducing the *ultimate* view of the essence within this Caribbean, Jamaican aesthetic. A male voicing the sound 'yeah man' opens the feature at the same time that a young male is seen plunging from a cliff face into a turquoise ocean. There is a golden sunset, a tropical lizard, palm tree, a shot of the Rasta (traditional Ethiopian) flag, a vividly colourful painting of a pineapple, a tourist woman in bikini on horseback, an elder Rastafari woman seated on a raft, the Jamaican Constabulary Marching Band, and a Rastafari priest making an address.

This is the sequence of shots that can be observed in the first ten seconds of the production. Bob Marley is the first Jamaican personality mentioned, and this is used almost as if to legitimise the prime minister's Jamaican pedigree. Further, the packaging of Jamaica's tourist product is done via the mention of Ian Fleming, J. P. Morgan, Bette Davis, Martin Luther King, Jr., and other famous persons who accessed Jamaica for leisure and recreation, inspiration, and as location for Hollywood feature films. Jamaican Blue Mountain coffee, Red Stripe beer, reggae music, and a range of experiences from the highest point to the internal rhythm of the population are presented through highlights of cricket, dominoes, exchanges with the people, and a tour of the Bob Marley Museum in Kingston (the only tourist feature of Kingston, the capital city, that was presented). Marley's widow Rita facilitated the tour of the Bob Marley Museum. That (approximately two-minute) feature ends with the prime minister and Rita singing Marley's famous song 'One Love.' Marley is in this moment arguably 'synonymous with Jamaica' and the first face that comes to mind with the mention of the island. This documentary, as has so many other features on Jamaica, draws on some of the internationally recognized symbols of 'Jamaican-ness,' the difference this time being that the guide was the prime minister of the country.

This 2005 feature tells an interesting story of how Jamaica has almost been simplified as being just a backdrop for a Bob Marley tale. Bob Marley has entered yet another phase of his journey beyond stardom, superstardom, and legend[6] and has become a part of the packaged presentations of Jamaica that can be made available to visitors and fans. Marley is perhaps more famous as an iconic representation of Jamaica than Blue Mountain coffee and reggae music combined. Since his passing in 1981, his estate has been steadily developed into a network of sites through which the Bob Marley magic can be nostalgically shared with visitors. There are four primary sites in Jamaica: Tuff Gong Recording Studio; the Bob Marley Museum; Trench Town Culture Yard; Marley's burial place at Nine Miles, St. Ann. Additionally, there is a well developed site in the United States (Bob Marley: Tribute to Freedom, at Universal Studios in Orlando, Florida). The Marley Museum tour is the major tourist attraction in the city of Kingston, and in this respect it outranks all other reggae music tourist sites and products one can experience anywhere. Marley, as a Rastafarian artist spreading reggae around the world, provides a convenient backdrop for exploring how it is that the Jamaican tourist product has been affected by the emergence of Rastafari in Jamaica.

AESTHETICS AND REGGAE TOURISM

Kwame Dawes (2004) is convinced of the idea of a *reggae aesthetic.* Dawes articulates and extends this aesthetic to an overarching Caribbean aesthetic,[7] which includes calypso and Indo-Caribbean artistic forms (p. 17). Reggae music, Dawes explains, contains principles of beauty 'that can help to define the arts that emerge from the world that has shaped reggae.' For Dawes, reggae presents a Jamaican bouquet of beauty: as he puts it, 'reggae is conceived first as Jamaican. Jamaicans own reggae and they regard it as part of the culturally defining mechanisms of their society—as Jamaican as Blue Mountain coffee. However, this allegory of commodification is only one part of reggae's nationalist credentials. Even more important to reggae's nationalist credentials is that reggae has helped to define Jamaica to itself' (Dawes 2004:31). Marley has therefore been a significant contributor to the construction and defining of reggae as an aesthetic.

It stands to reason that Jamaican tourism can also be argued to have experienced a reggae transition and interpretation increasingly since the mid-1970s and more especially since the death of Bob Marley. We would further argue that Jamaica and those marketing and characterizing Jamaica's unique beauty and cultural quality exploit the Rastafari and reggae contributions, conveniently to distinguish the signal nature of the island's aesthetic. In the feature 'Jamaica the Ultimate Tour,' the Rastafari woman and man are the only persons profiled with the exception of the female tourist (in the opening ten seconds). The traditional Ethiopian flag generally described as the Rastafari flag is the only one that is highlighted in a feature that has the Jamaican head of state as guide. Rastafari elder and teacher Mortimo Planno (1998) in his 'Polite Violence' Lecture explained this consumption of the movement's image as a part of an exploitation of the brethren's exoticism, and he claims that official slogans now invite visitors to 'come to Jamaica and watch the Rasta-man walk.' He further explained, 'they become attracted to our fascination.' This attraction to the Rastafari 'fascination' that emerged with the routinisation of the movement in the 1970s resulted in what he described as 'a deprogramme of the back-to-Africa objective.'[8] Back to Africa, up until the late 1960s, before the emergence of Bob Marley as a leading Rastafari artist, was the central philosophical and political objective of the movement, with Ethiopia as its key regional focus inspired and guided by the leadership of its Emperor Haile Selassie I.

As early as the late 1960s (about the same time as the emergence of reggae music), Leonard Barrett, famous scholar of the movement, argued that the Rastafari movement would potentially have deleterious effects on the population's economy and, in particular, on the tourist market (Barrett 1968:176–77, 189). This was not dissimilar to that opinion held by the island's first prime minister, Alexander Bustamante, who in 1963 waged a manhunt to capture some of the brethren involved in an incident at Coral Gardens in St. James, the island's tourist capital.[9] This placed an ugly stigma on those within the movement as criminals and murderers, as being antisocial as well as social misfits, and the idea of back to Africa as escapism. To therefore argue for a reggae/Rastafari aesthetic as a type of Jamaican tourist product is also another irony of history because the Rastafari were seen to be damaging the product immeasurably in their revolutionary move away from traditionally held Jamaican identity, politics, and religion.

The Rastafari movement's people are among the most significant cultural community in Jamaica today. This stems largely from the interest of 'outsiders' or 'others' in observing/witnessing and participating/experiencing the Rastafari philosophy and *livity* (or way of life) firsthand. Such interest originated within the international scholarship from the early 1950s, and the movement's 'fascination' had attracted local press coverage from as early as the 1930s. By the end of the first four decades of the Rastafari movement's development, researchers from North America, the United Kingdom, and the Caribbean had written studies of the local phenomenon (Niaah 2006). Gradually research interest in the Rastafari has expanded to Western Europe, Asia, South America, and Africa throughout the decades of the 1970s to the present. It is perhaps due to the growth and development of Rastafari and reggae adherents in these regions why researchers from such regions became interested in making connections with the source of the Rastafari existence. Interest in Rastafari and reggae are often driven by the encounter of Rastafari in unusual or unexpected places such as Japan, New Zealand, Guam, the Grand Canyon, Finland, or Italy. To this extent the United Kingdom is among the earliest sites of Rastafari export. Jamaica, Asia, Africa, and South America among the more recent regions where the movement's influence has developed.

The global reach of the Rastafari, however, cannot be accredited to the academic survey and scrutiny (tourism of sorts) but more so from the influence of reggae music which spread the Rastafari message as well as the general overarching themes of reggae as protest music to the world. Dawes explains it as follows: 'Reggae's arrival constitutes a pivotal and defining historical moment in the evolution of a West Indian aesthetic, an important stage that stands beside a series of other important stages' (Dawes 2004:14).

Reggae within its first decade had formulated itself as an idiosyncratic fusion of urban/rural, local/foreign, poor/rich aesthetic of sound, locally and internationally. Dawes believes that the aesthetic of reggae has subsequently dominated the information coding and presentation of Jamaica and the Caribbean experience in general.[10]

Prior to the emergence of reggae and its advancement as the leading sound of Jamaica within a decade, calypso was the assumed sound of the Caribbean, even when the music being played was not calypso. Neely (2006:1) provides us with some insights of what might be described as a 'calypso aesthetic' by indicating that the 'Jamaican caplysonian' was really a 'mento' musician who was featured in rural farming communities and had become incorporated into the live music entertainment package of the tourist industry from as early as 1930. Outside of the increased popularity of sound systems to provide musical entertainment in tourist resort communities, mento as a music form in Jamaica became increasingly replaced by ska, rock-steady, and eventually reggae (between 1958 and 1968) and dancehall music as the emerging sound of Jamaica. In the same way that calypso provided an overarching label for the Caribbean, a feature which is still part of the Caribbean aesthetic in the tourist fantasy, so too is this reggae aesthetic becoming an overarching feature of the Caribbean, often to the surprise of Jamaican tourists who arrive in other islands and find the same dominance of reggae in public media as in Jamaica.

There is a preconception in the tourist vision of the Caribbean and Jamaica in particular, especially among visitors who originate from temperate climates and anticipate a tropical reprieve through their island vacation. Among the anticipated symbols media images utilize are yellow/red tropical sun; generally the presence of vibrant colours (flowers, birds, and dress); white-sand beaches; lush mountain ranges; bacchanalian living; warm and friendly service; fit/healthy, agile, and skillful bodies in rhythmic motion; beautiful smiling faces (some derived from racial mixture); spicy foods and exotic fruits; a playground in 'paradise'; sensual and sensuous. Jamaica is routinely marketed in magazines and on television with these stereotypes, especially advertisements from the Jamaican Tourist Board's North American campaign. Added to this are the features of the Rastafari movement as an indigenous, exotic (or outrageous) addition, coupled with the sound of reggae music.[11] To this degree (though the phenomenon is no longer unique to Jamaica), Rastafari and reggae become additional components within the Jamaican tourism product and its aesthetic, to the extent that Blue Mountain coffee or Jamaican rum also offer these additional features of interest and constitute a part of the aesthetic.

To speak of a reggae aesthetic in Jamaican tourism is to therefore pronounce a mood and state of being within a history of natural tropical splendor, capped by the active agency of Africans and other nationalities negotiating the exploitation of this space. The reggae aesthetic speaks to the Jamaican translation of the experience in fashioning something beautiful from this Caribbean space. This reggae aesthetic in the Jamaican tourism product is demonstrated at its highest level through the legacy of Bob Marley. Bob Marley's character is the personification of the aesthetic. In the way that this aesthetic is originating from the Jamaican people, it exercises a counterhegemonic understanding of beauty (see Meeks 2000), especially given the historical background from which the official society has operated. Jennings (2000) describes this feature of the Rastafari contribution as an oral test that provides 'a hermeneutics of Babylon.' This leads to an aesthetic that is exploited successfully and works for Jamaica as a tourist attraction despite its recent international reputation of violence and civil unrest. To some extent the aesthetic accommodates those tumultuous features of the society and exploits these as part of the island's 'appeal.' This is endorsed by the direction of the recent films being produced out of Jamaica and their portrayal of violence.

In a real sense the ability to discern a sense of the reggae aesthetic genuinely can be had through the Jamaican film classic *The Harder They Come.* Dawes (2004: 28–29)identifies as a starting reference to the essence of his concept. According to Dawes, that film sets a ceiling on Jamaican filmmaking that no other has yet been able to capture. Films such as *Rockers, Country Man,* and *Children of Babylon* all experiment with this reggae aesthetic, exploiting and exploring all the components from the island's uniqueness, often with the Rastafari and reggae music as central features. Such features do exploit the themes of being in the Caribbean, but there is always the separate identity that the placement of Rastafari, Bob Marley, and reggae in these advertisements and features afford. Indeed, Bob Marley's 'One Love' song has been used as the island's tourist jingle for over a decade. It can also be seen within the promotional footage that often sport the image of dreadlocks (often dreadlocked children with White tourist children), and the vivid colour and beauty transmitted by the movement's culture.

THE BOB MARLEY LEGACY

One of the narratives presented about Bob Marley comes through an interpretation of Rastafari elder and Marley's Rastafari mentor, Mortimo Planno. In his assessment,[12] there is reason for paralleling the impact and contribution Marley has brought with an amalgam (or syncretic/indigenous version) of the tradition of Christ and his purpose. As an illustrious Jamaican son, Marley's expressed life work has glorified his spiritual father who lived in Ethiopia. To Planno there was a familiar story (an old Bible story) which was being retold in Bob Marley. This was now being confirmed by the honours and accolades visited on Marley's memory.

Bob Marley's estate continues to earn millions of dollars every year. There are also arguments surrounding how his creative work encapsulated in a musical empire now estimated to value in excess of US$100 million is to be distributed. Rita Marley in this regard took on a role similar to Amy Jacques Garvey by commandeering Marley's memory into a larger-than-life presentation. Included within the iconography and inscription of Marley's posthumous image are a marble mausoleum as a special resting place for his remains at his birthplace; a museum at his former residence; a museum in Trench Town, the place of his roots; statues to his honour (we are aware of four statues);[13] and the national honour of Order of Merit of Jamaica.[14] The government of Jamaica has ordered gold and silver coins to be minted in his honour on two occasions (his fiftieth and sixtieth birthday anniversaries). There are a plethora of references to Marley in song and countless Bob Marley cover versions. Books and printed matter[15] with the Marley story or the exploitation of his likeness are produced yearly, so much so that the estate of the artist has chosen to take legal measures to protect the 'face' and 'name.'[16]

There has been a careful construction of the Bob Marley story as well as Marley myths. Marley is said to have been a special child, a clairvoyant child, and palm reader. He is also said to have been lost at the age of seven in Kingston, almost similar to the idea of Christ being lost, separated from his parents as a youth, and later found among the scholars of the time. At the Bob Marley Museum the packaging of the Marley mystique opens with a statue of him guarded by a pair of lion statues (200 years old) imported from West Africa, placed at his feet. Visitors can share in an hour-long tour that takes them on a sort of *pilgrimage* (as some considered it)[17] through the daily routine of Marley in his domestic and work space, including a view of the bullet holes in the walls where he was shot and escaped with his life.[18] Album covers, his gold and platinum music awards, and a film are among the things that can be viewed in this space, sometimes sealed by a glimpse of his widow or one of his children traversing the grounds. The Marley Museum can easily be counted as Jamaica's number one reggae tourist destination since there is no other enterprise that can deliver this authentic reggae tour.

This chapter does not examine the merchandising of Marley in buttons, T-shirts,[19] clothing, and memorabilia (including candles, incense, cigarettes, cigarette paper, necklaces, and pins) within and without the tourist sector, the celebration of Marley in places such as New Zealand, or the local and national events built around the name Bob Marley and the celebration of his life. What is here examined is the importance of the Marley product to visitors to Jamaica, a legacy that many Jamaicans took for granted until faced with the idea of moving Marley's remains when this story was circulated in the media a few years ago. This spurred great public outcry and a vibrant media-driven reaction to the possibility of the family removing Bob's remains for final burial in Ethiopia.[20]

Figure 3.1 Statue of Bob Marley at the National Stadium, in Kingston.

A NEW PRISM: REGGAE AND RASTA AT
THE BOB MARLEY MUSEUM

The museum is a permanent site where Bob Marley fans can celebrate his life and work, and for a handful, visits acquire ritual significance through repetition. And even for a larger cross-section of the visitors, this tour would constitute the most comprehensive and legitimate view of an authentic reggae aesthetic. Outside of just a display of Marley memorabilia, the museum site also accommodates Jamaican cuisine, Rastafari *ital* food[21] and natural juices, and a hairdressing parlour that specializes in braiding and dreadlocking. There is also a clothing store as well as a gift/souvenir store. The former specializes in the Bob Marley trademark shoes, T-shirts, and other garments. The souvenir store supplies handicraft from Africa along with Jamaican product and other goods such as posters, buttons, pins, earrings, stools, carvings, fabric, oils representing Rastafari lifestyle, and symbols of choice. The site is therefore one which is highly self-contained (or all inclusive) and provides a fruitful space for engaging with the underresearched phenomenon of what is being described as a reggae/Rasta tourism product.

In research conducted between 2003 and 2005, a survey of visitors and recordings in the museum's visitors' book was used to classify visitors, the reason for their visit, and international perspectives on Marley. It is estimated that the Bob Marley Museum attracts about 30,000 visitors each year. These visitors come from all around the world, and their comments about the experience at the Bob Marley Museum give voice to the experiential dimension of the reggae aesthetic. Visitors' region of origin is shown in Table 3.1.

The Museum can be recognized as a financially viable attraction in Kingston: visitors pay US$10 for the tour leading to estimated revenue of US$300,000 per annum based on entrance fees. The site is also attractive to an international clientele, in particular the European tourists (including those from the United Kingdom), who account for almost 50 percent of the

Table 3.1 Visitors' Region of Origin (July/August 2003)

Country	Number of Visitors
Europe	295
North America	241 (of which USA—218)
United Kingdom	178
Caribbean	100 (of which Jamaica—58)
Asia	21 (of which Japan—13)
Australia /New Zealand	20
Africa	14
South & Central America	13
Middle East	3
Total	885

Source: Visitors' Book Survey 2003

sample of visitors in the preceding table. The visitors from the Caribbean (including Jamaica) ranked fourth overall, out of nine regional categories, by way of visitor numbers during the survey period. Visitor interest/numbers reflect the general regional emphasis of the tourist market by way of country of origin. However, as noted previously, Europe combined factored more than it did by way of ratio of its overall visitor arrival in the island and the percentage of those who participated in the museum tour.

Visitors to the museum, as at almost every other music tourism site, included those on a package tour; those who came because Bob Marley has iconic status to them; and others because they were told it was the thing to do. Despite the motivation for visiting the museum, tourists generally enjoyed the experience. Their comments recorded in the visitors' book at the Museum ranged from the acknowledgment of Marley as a 'revolutionary who had risen from the bottom of the society, as one who 'brings people together,' the 'ambassador of world music,' or a spokesman for 'peace and solidarity.' One visitor/fan believed that the museum was a 'must' for all reggae fans visiting Marley's country. Such reggae/Marley fans all tended to mention owning a copy of the Marley compilation album 'Legend.' There was a sense that the Marley story was also identified as 'part of Jamaica's history.' Overall, the ideas of the Japanese tourists perhaps best capture the essence of the comments in the visitors' book. One noted that 'his wisdom was very rare' and he was a man of 'great respect.' For some visitors, Marley was important enough for all the Marley attractions to be covered on their trip. This would include visits to Culture Yard in Trench Town, the Tuff Gong recording studio on Spanish Town Road, and Marley's burial spot at Nine Miles in St. Ann. One truly die-hard fan indicated having every recording and had first listened to his music in the 1960s as a youth in England. For him this was a pilgrimage to pay respect to the king since Marley meant 'one love and the unity of the African peoples.' Many say that they derive inspirational experiences from the tour.

The museum seems to satisfy a sense of civic pride among the Jamaicans who live abroad, and hence they constituted more than 50 percent of the Caribbean nationals who visit the site. For many of them, Marley's life represents a type of Jamaican story and experience about success against the odds. But he also represented the documentation and ability to share in a tangible aspect of the Jamaican marginal culture that is usually not available in that type of packaging (i.e., in a museum). For the expatriate Jamaicans, various aspects of Marley's life had particular resonance, and his songs helped them through their overseas experiences. One returning Jamaican would have liked the museum to have more about Rastafarian culture. As it stands, the Bob Marley museum is the closest thing to a Rastafari archival repository in Jamaica. A Jamaican expatriate who had taken his (foreign-born) daughter remarked that Marley 'gave pride as a Jamaican and his messages gave Black people respect in contrast to

Figure 3.2 Photograph of the Bob Marley Museum, taken at the gate of the Bob Marley Museum.

American music.' The museum therefore satisfied a sense of fulfilment about a Jamaican heritage.

Visitors' books at many locations are usually replete with frivolous sensational one-liners as tributes to the varying sentiments they seek to express. At the Bob Marley Museum, out of some 900 entries received per month from visitors, none of these were seemingly 'frivolous' or insincere. Below are some of the sentiments expressed by those who visited and signed the book provided by the museum: 'Bob we love you'; 'awesome'; 'enlightening and sad'; 'Jah Rastafari'; 'Trust in God'; 'thanks'; 'peace, love, unity, respect'; 'great to be here'; and 'best part of my visit to Jamaica.' Despite all of this, there are visitors who are not impressed by the extent to which the Bob Marley brand has been commercially exploited, and hence one of the respondents mentioned the commercial nature of the museum, capped in the comment 'nice but commercial.' Reggae and the related culture that the music emanates from are often perceived to have 'sold out' from their core identities as a result of commercial success. The visitors who come to the location generally capture the religio-political character and significance of the Rastafari movement and reggae as a musical form. One visitor remarked that the museum was like a 'sacred church and rebel sanctuary combined.' This sense is even more underscored when one visits Nine Miles, the location of the Marley mausoleum, which is built in the shape of a traditional Ethiopian Orthodox church and includes an area where pilgrims can pray. In the various perceptions there were no evident differences between those who had come from Europe, Britain, or North America.

It should also be noted that many of the visitors to the Bob Marley Museum indicated that they had chosen Jamaica as a destination because of Bob Marley, and among most of those who did, the decision to visit the site was made before arrival. An Australian who toured the museum one day after arrival on the island concluded that the tour was the best start imaginable for his trip to Jamaica. Another visitor, perhaps in trying to express similar appreciation, reported that Bob Marley 'goes with the culture and climate of Jamaica.' A Zimbabwean medical doctor living in London, who met Bob Marley in 1980, indicated that Marley meant 'Good music, interesting lyrics, [we] play it during our operating sessions . . .' It is a forty-one-year-old teacher from London who seems most passionate about the Bob Marley phenomenon. She stated:

> Bob Marley's image did not mean much to me. But his words and what he tried to do has really helped encourage black children, and people around the world, to see that you change people's view of their social and economical background. Bob Marley's songs and life story should be part of the curriculum, songs are wonderful for motivating people especially young people who are disenchanted.

The museum, opened in 1987 to visitors, is a relatively suburban (uptown) site. However, for those visitors who are interested in the earlier roots of Marley, in the year 2000, his former 'government yard' residence was officially opened as a tourist attraction. This reportedly came after streams of tourists visited the site annually to share in the creative space from which many of Marley's songs had developed. It is also reported that tourists had previously come into the 'old yard' and camped for days at a time to pick up the 'vibes.' As one of the tour guides indicated, Culture Yard in Trench Town offers the 'poverty living experience' of Marley. Since Culture Yard opened, it has attracted a steady stream of international visitors and tourists. It has also seen various different contributions from international and local sponsors to complete what is expected to be a compelling reconstruction and (tourist) expansion of Marley's Trench Town days and restoration of downtown Kingston by extension.

CONCLUSION: TOURISM AND THE PEOPLES' LIBERATION

Tourism is now a major focus of not only Caribbean but also global economies geared at meeting the needs of holiday makers who wish to purchase a concept(s) of recreation and leisure, usually away from home. In the Caribbean there is increasingly the construction of a reggae aesthetic generally and a music tourism product more specifically. With this in mind, Jamaica and Bob Marley enter the marketplace with decided advantages for marketing this reggae aesthetic, and the Jamaican society has exploited this while ambivalent officially about the value of Bob Marley's contribution to the country. Despite the overarching official attitude, Bob Marley and his medium of reggae are the essence of the Jamaican tourist product, and it is what visitors connect with prior to their arrival. This aesthetic has replaced the previous calypso/bacchanalian/banana-producing image present before the emergence of reggae as a world music. The product of the consciousness emerging from reggae is a religio-political translation of the Jamaican society to the world as well as to itself. Most interesting is the irony of the elevation (demotion, some may argue) of the rebel Marley to the position of one of the chief tourist attractions and most highly acclaimed citizens. And if Jamaicans at home don't often recognize this, the museum is visionary in its product, and international visitors generally have a strong consciousness of the place of Marley in their experience of Jamaica.

NOTES

1. This chapter was developed from ongoing research into the celebration of Bob Marley. See J. Niaah, 'The Canonisation of Bob Marley,' paper presented at the 11th Annual International Conference of the International Society for African Philosophy and Studies (ISAPS), Enugu, Nigeria, March 2005, and J. Connell and Stanley S. Niaah, 'Remembering the Fire: Tourism at the Bob Marley Museum, Kingston, Jamaica,' *Caribbean Geographer* 14(1) (forthcoming).
2. See S. Wynter, 'We Know Where We Are From': The Politics of Culture from Myal to Marley, unpublished (Houston Conference), in the CLR James Collection at Brown University, 1977, for a discussion of Marley in a continuum of resistance practices stemming from Myal. See B. Chevannes, 'Speech at the Rastafari Cultural Gathering in Commemoration of the Coral Gardens Incident,' Sam Sharpe Square, Montego Bay, Jamaica, unpublished, transcribed by Michael Kuelker, April 2, 1999a, for discussion of Rastafari as memory of Jamaica.
3. For more details of the Rastafari movement, see H. Campbell, *Rastafari and Resistance: From Marcus Garvey to Walter Rodney,* Africa: World Press, Inc., [1987] 1994, B. Chevannes, *Rastafari Roots and Ideology,* The Press: University of the West Indies, 1994, and M. G. Smith et al., *The Report on the Ras Tafari Movement of Kingston, Jamaica,* ISER, 1960.
4. See R. Hill, *Dread History: Leonard P. Howell and Millenarian Visions in the Early Rastafarian Religion,* Chicago: Research Association School, Times Publications/Frontline Distribution Int'l Inc., 2001, for full discussion of Howell and other early pioneers of the movement.
5. See E. Hopkins, 'The Nyabinghi Cult of Southern Uganda,' in R. Rotberg, ed., *Protest and Power in Black Africa,* Oxford: Oxford University Press, 1970, for a discussion of the *Nyabinghi* origin, and H. Campbell, *Rastafari and Resistance: From Marcus Garvey to Walter Rodney,* Africa: World Press, Inc., [1987] 1994, for data related to the Nyabinghi in Jamaica.
   Also see E. Leib, 'Churchical Chants of the Nyahbinghi,' Heartbeat CD, Poli-Rhythm, Ltd., Cambridge, MA, 1997, for audio tracks of the music.

Y. S. Nagashima, *Rastafarian Music in Contemporary Jamaica: A Study of Socio-Religious Music of the Rastafarian Music in Jamaica,* Tokyo: Institute for the study of Languages & Cultures of Asia, 1984, is also useful as well. As pertaining to *burru* music, see **M. G.** Smith et al., *The Report on the Ras Tafari Movement of Kingston, Jamaica,* ISER, 1960, as well as V. Reckord, 'Rastafarian Music—An Introductory Study,' in *Jamaica Journal* 10 (1,2), 1976; V. Reckord, 'From Burru Drums to Reggae Ridims: The Evolution of Rasta Music,' in N. S. Murrell, W. D. Spencer, and A. A. McFarlane, eds., *Chanting Down Babylon: The Rastafari Reader,* Temple University, 1998.

6. B. Chevannes, 'Between the Living and the Dead: The Apotheosis of the Rastafari Hero,' in Pulis, ed., *Religion, Diaspora, and Cultural Identity,* 1999b. This speaks of apotheosis to track Marley even beyond this category to hero celebration.

7. M. Anderson, *'Cahuita Gone, Cahuita Gone': Struggles over Place on the Caribbean Coast of Costa Rica,* unpublished doctoral dissertation, University of Cambridge, 2002, adds dimension to this idea through what she describes as Caribbeanisation of a tourist locale in Central America. This essentially meant inclusion of reggae and 'Rasta' visual aesthetics as a clear part of the tourist product to give a feeling of being 'just like Jamaica' (p. 93).

8. By deprogramme of the back-to-Africa objective, Planno is referring to the move towards a more sedentary posture within the West, which some attribute to increased commercial success within the movement through reggae music (among other things). For access to the abovementioned lecture, see M. Planno, 'Polite Violence: The lecture,' in Mortimo Planno collection, Library of the Spoken Word, Radio Education Unit (LSW,REU), University of the West Indies, Mona, 1998.

9. For discussions of the Coral Gardens incident, see F. Van Dijk, *Rastafari and Jamaican Society,* Utrecht, the Netherlands: ISOR, 1993.

10. Dawes's view extends into the realm of the music as sensuous as well as sensual relating to the Jamaican body and imagination. Though there is some knowledge of other connections that the image of Rastafari and tourism together hold as it relates to sexual service and drug suppliers (see, for example, M. Anderson, *'Cahuita Gone, Cahuita Gone': Struggles over Place on the Caribbean Coast of Costa Rica,* unpublished doctoral dissertation, University of Cambridge, 2002).

11. See E. Brodber, 'Black Consciousness and Popular Music in Jamaica in the 1960s and 1970s,' *Caribbean Quarterly* 31 (2), 1985, for a discussion of how the growing influence of the Rastafari helped to shape her own Black consciousness and that of others in the society.

12. See M. Planno, 'Polite Violence: The lecture,' in Mortimo Planno collection, Library of the Spoken Word, Radio Education Unit (LSW, REU), University of the West Indies, Mona, 1998.

13. The attempt to construct his likeness sparked a huge public debate (after its presentation) about the interpretation of the artist and whether the image looked like Bob Marley.

14. This order is still virtually under debate as there are some sections of the population who believe that Marley is deserving of the honour of Nation Hero, the highest national honour.

15. Perhaps a guide to the existing works featuring Marley is provided by the almost 3 million references to his life and work on the Internet to date.

16. See B. S. M. Hylton, and P. Goldson, 'The New Tort of Appropriation of Personality: Protecting Bob Marley's Face,' *Cambridge Law Journal* 55 (1), 1996:55–64.

17. J. Connell, and S. Stanley Niaah, 'Remembering the Fire: Tourism at the Bob Marley Museum, Kingston, Jamaica,' *Caribbean Geographer* 14(1) (forthcoming) introduce this idea in referring to the visitors to music tourism sites for whom such visits constitute a journey to a sacred place (sometimes dedicated to someone venerated by such guests).
18. In December 1976, Bob Marley was shot in the hand and his wife Rita in the head. Days later he appeared on the Smile Jamaica concert and performed indicating clearly that he would not be intimidated by political activism and threats to cease his work of cultural politics. Five months later, while playing in a friendly international soccer match in Paris, Bob Marley's big toe was spiked (mashed) and this later proved to be a fatal blow resulting in the contraction of melanoma cancer.
19. T-shirts are an interesting part of the reggae aesthetic. Marley's image is the most dominant. Images of the Ethiopian Emperor Haile Selassie I, Marcus Garvey, and Peter Tosh are among the classic images.
20. Among the Rastafari, many wish to return to Africa, Ethiopia in particular, and leave behind the pollution of Babylon. This vision comes as part of a long tradition of back-to-Africa advocacy from the time of the maroons.
21. *Ital* food refers to what is generally considered as the diet of the Rastafarian brethren, *ital* considered to be derived from 'natural' or 'vital,' usually in reference to organic foods and mostly vegetarian menus.

REFERENCES

Anderson, M. 2002. 'Cahuita gone, Cahuita gone': Struggles over place on the Caribbean coast of Costa Rica. Unpublished doctoral dissertation, University of Cambridge, Cambridge.

Barrett, L. 1968. *The Rastafarians: The creadlocks of Jamaica.* Kingston: Sangster's/ Heinemann Educational Books.

Brodber, E. 1985. Black consciousness and popular music in Jamaica in the 1960s and 1970s. *Caribbean Quarterly* 31 (2).

Campbell, H. [1987] 1994. *Rastafari and resistance: From Marcus Garvey to Walter Rodney.* Trenton, New Jersey: Africa World Press.

Chevannes, B. 1993. The Rastafari and the urban youth. In *Perspectives on Jamaica in the seventies,* C. Stone and A. Brown, eds. Jamaica: Kingston Publishers pp 392–422.

———. 1994. *Rastafari roots and ideology.* Kingston: University of the West Indies Press.

———. 1999. Speech at the Rastafari Cultural Gathering in commemoration of the Coral Gardens Incident, Sam Sharpe Square, Montego Bay, Jamaica. Transcribed by Michael Kuelker.

Connell, J., and S. Stanley Niaah. Forthcoming. Remembering the fire: Tourism at the Bob Marley Museum, Kingston, Jamaica. *Caribbean Geographer* 14 (1).

Dawes, K. 2004. *Natural mysticism: Towards a new reggae aesthetic in Caribbean writing.* UK: London: Peepal Tree Press.

Hill, R. 2001. *Dread history: Leonard P. Howell and millenarian visions in the early Rastafarian religion.* Chicago: Research Association School Times Publications/ Frontline Distribution Int'l Inc.

Hopkins, E. 1970. The Nyabinghi cult of Southern Uganda. In *Protest and Power in Black Africa,* R. Rotberg, ed. Oxford: Oxford University Press.

Hylton, B. S. M., and P. Goldson. 1996. The new tort of appropriation of personality: Protecting Bob Marley's face. *Cambridge Law Journal* 55 (1):55–64.

Jennings, S. 2000. *The hermeneutics of 'Babylon' in selected pieces of Jamaican popular music of the 1970s.* Paper read at Cultural Studies Colloquium.

Leib, E. 1997. *Churchical chants of the Nyahbinghi.* Cambridge, MA: Heartbeat CD, Poli-Rhythm.

Meeks, B. 2000. *Narratives of resistance: Jamaica, Trinidad, the Caribbean.* Kingston: The Press.

Mekfet, T. 1993. *Christopher Columbus and Rastafari: Ironies of history and other reflections on the symbol of Rastafari.* St. Ann, Jamaica: self-published

Nagashima, Y. S. 1984. *Rastafarian music in contemporary Jamaica: A study of socio-religious music of the Rastafarian music in Jamaica.* Tokyo: Institute for the Study of Languages and Cultures of Asia.

Neely, D. T. 2006. Tourism is our business: The changed role of calypso in Jamaica. In *The Barry Chevannes Conference.* Kingston: UWI Mona.

Niaah, J. 2005. *The canonisation of Bob Marley.* Paper read at 11th Annual International Conference of the International Society for African Philosophy and Studies (ISAPS), at Enugu, Nigeria.

———. 2006. Rasta teacher: Leadership, pedagogy and the new faculty of interpretation. Unpublished doctoral dissertation, University of the West Indies, Mona, Kingston.

Planno, M. 1998. *Polite violence: The lecture.* Mona: Radio Education Unit (LSW, REU), University of the West Indies.

Reckord, V. 1976. Rastafarian music—An introductory study. *Jamaica Journal* 10 (1/2).

———. 1998. From burru drums to reggae ridims: The evolution of Rasta music. In *Chanting down Babylon: The Rastafari reader,* N. S. Murrell, W. Spencer, and A. McFarlane, eds. pp 231–252. Kingston: Ian Randle Publishers.

Smith, M. G. 1960. The report on the Ras Tafari movement of Kingston, Jamaica. Kingston: ISER.

Van Dijk, F. 1993. *Rastafari and Jamaican society.* Utrecht, the Netherlands: ISOR.

Wynter, S. 1977. We know where we are from: The politics of culture from Myal to Marley. Unpublished (Houston Conference), in the CLR James Collection: Brown University.

4 Jamaican Vinyl Tourism
A Niche within a Niche

Douglas Webster

INTRODUCTION

The Caribbean island of Jamaica boasts a long and distinguished history as a tourist destination, but essentially so, as what might be called a "three S" (sun, sea, and sand) target. Given the island's geological endowment, favourable tropical location, and proximity to two of the world's main traditional centres of tourism demand, the United States and Canada, Jamaica continues to be well positioned to deliver the traditional tourism package.

Since the outset of the twenty-first century, with growing competitiveness in the global tourism market, a modification of that narrow perspective is being sought as a matter of national policy (Government of Jamaica 10 Year Tourism Master Plan 2001). Now there are active efforts afoot to inform the wider world that as a tourism hub there is so much more to Jamaica than white sand beaches, the foaming blue Caribbean Sea, and ninety-degree (F) temperatures. Nature-based tourism, for instance, will entail tourist forays into the relatively unexplored, lush interior of this very mountainous island to allow for their experiencing Jamaica's bewildering biodiversity, microclimates, and diverse terrain.

THE CULTURAL DIMENSION

But another essential endowment with which Jamaica is blessed in abundance is that priceless intangible called culture, that distinct manifestation of a people's unique national mindset or psyche via expressions such as dance, music, and the concretisation of the abstract in the form of written poetry, pottery, and sculpture. That, coupled with a long, varied, and fascinating history (with a rich admixture of the good, the bad, and the indifferent), makes for what might be termed a distinct *heritage* or that strong cultural legacy that appeals to students of the exotic, hard-nosed researchers of sociology and "folkways," and the ordinary Joe wishing for a brief "psychic" respite from the daily grind of the typical First World city.

REGGAE TOURISM

Evidently, heritage tourism or cultural tourism is potentially one important component of Jamaica's overall tourism product, but it too has its subgenres. Let us posit reggae tourism to be one such subcategory. But in and of itself, reggae tourism is quite wide, entailing, inter alia, visits to significant historical sites including late reggae icon Bob Marley's birthplace, a veritable shrine, and attendance at major music festivals including the recently revived Reggae Sunsplash, and Reggae Sumfest. It also incorporates an element of business tourism, if you will, with musicians visiting to record in Jamaica and "catch the vibes" that they feel reside in their purest form only in the birthplace of reggae.

DEFINING REGGAE

Undoubtedly *the* single most celebrated aspect of Jamaican culture is the *artistic expression* known as reggae, which debatably has become the most popular Third World music form. Broadly defined, reggae music refers to Jamaican popular music from the early nineteen sixties through the nineteen nineties. A distinction is now being made between reggae and dance hall. The subgenres of reggae, in chronological order or their *advents,* include:

- Ska; with its choppy up-tempo syncopation and room for improvisation, this was to all intents and purposes Jamaica's jazz. Ska was something of a musical amalgam though, having its roots in indigenous Jamaican idioms such as mento, combined with the unmistakeable influence of New Orleans–based rhythm and blues and, of course, the original jazz of African America
- Rocksteady; tempo-wise, a polar opposite to ska (almost a psychological overcompensation for the frenetic pace of its predecessor), relying on slow pulsating rhythm and intricate guitar licks, facilitating controlled close harmony singing. And by and large, gone was the brass as dominant solo instrument and vehicle of melody
- Revive reggae (a curious misnomer for an early manifestation of reggae, then spelt "r-e-g-g-a-y" in some quarters); representing another tempo reversal, some would argue a correction, with the pace being stepped up to satisfy the energetic dancing feet of Jamaicans. For that reason, the author's terminology for that subgenre is *tripping* reggae
- Reggae (which entailed something of a slowing down of the tempo from its early "revive" form); with its pronounced bass line and "one-drop" drumming pattern, a perfect midtempo vehicle for the twinning of rhythm and melody. In its pure form, it carried throughout the decade of the seventies until drum machines, synthesizers, and other computerised effects began to make their presence felt

- Dance hall, defined in some quarters as distinct from reggae, is the province of the deejay, a chanter/rapper of sorts rather than a true singer. It is digitally constructed, rhythmically rudimentary, and more often than not, lyrically raw and in your face. This is the current day manifestation of popular Jamaican musical expression.

THE MOTIVATION

This study is an attempt to give formal expression to working knowledge, some anecdotal, some firsthand, garnered from over thirty years of close association with the Jamaican record retail scene in both its formal and informal manifestations. One development noted over the years was a growing international market for old Jamaican records. It was appreciated that overseas visitors sought these records, but it was assumed at first that record purchases were by culturally adventurous visitors and incidental to the traditional "three S" reason for making Jamaica a vacation destination.

Subsequently, the concept of the vinyl tourist came about intuitively from hearing about and, in the author's capacity as a record collector, sometimes having to compete with, persons who were identified as being in Jamaica for no other reason than to purchase old Jamaican records. That was felt to be worthy of investigation.

Looking now at the issue of buyer motivation, only a die-hard record collector can fully appreciate the "buzz" achieved from acquiring decades-old vinyl platters in excellent working condition with relatively unblemished (and original) labels. To vinyl aficionados there is intrinsic value to owning an original vinyl recording of a rated work by a legendary artiste or even an artistically commendable but commercially unsuccessful work from a little-known artiste, the latter in particular making for that much-sought-after rarity premium. Despite the presence of pops and clicks from records, there are those audiophiles who insist that no digital medium (compact disc, DVD-audio), despite its clinically clean reproduction, provides the "warmth" and richness of the output from a vinyl record. Such factors help drive consistent demand.

This demand on the part of collectors creates a market, a need to be filled, so considerable money values are involved, and no doubt some collectors acquire vinyl records as a form of investment. Beyond that, though, is the factor of owning a record for the sake of owning it; this accords bragging rights to serious collectors who by nature can be a fiercely competitive breed.

It is no surprise to the author, therefore, that persons will travel thousands of miles and roam about, even at some risk, in a relatively unfamiliar country to acquire vinyl for themselves or, more typically, on behalf of others with a driving passion for ownership of these historically significant products.

The Methodology

This chapter was inspired out of keen interest and observation and grew out of a desire to make somewhat more structured enquiry and conclusions through interviews with "players" on the vintage vinyl scene. There were some a priori assumptions about the landscape and behavioural factors regarding the supply side, but as noted in the study at least one of these assumptions was challenged by evidence.

The empirical aspect of the study rests on six elite interviews, four with record dealers and two with vinyl tourists. Thus they represent both the demand and supply sides of the vintage vinyl market and offer a glimpse into the dynamics of that market, which obviously functions well, albeit in a highly informal fashion. On the dealer side, interviews were conducted with those who saw the birth of the vinyl-tourist movement and those currently "on the road" in the thick of things presiding over what may be the maturation of that subniche of the tourism market.

On the demand side, a representative from the lead country, Japan, was canvassed as was one dually representing Latin America and Europe.

VINYL TOURISM: THE PRACTITIONER AND THE PRODUCT

This study seeks to define and substantiate the existence of a subset of reggae tourism which, with no state initiative or encouragement from tourism industry interests, assumed sufficient significance to be taken seriously and be subject to research and analysis. This is vinyl tourism, a specialist tourism subproduct that constitutes visits to Jamaica for the sole and express purpose of seeking and purchasing Jamaican vinyl records, especially those of the sixties and seventies in the ska, rocksteady, and nascent reggae idioms. Records of the nineteen eighties and early to mid-nineties are now also coming into demand, as are non-Jamaican records short of supply worldwide. But a valuable by-product gained is knowledge, including market intelligence to maximize chances for success on future buying excursions.

PROFILE OF THE VINYL TOURIST

What is the profile of the typical vinyl tourist? Given his mission (and the "his" is not generic; the vinyl tourist is, to all intents and purposes, invariably male), the vinyl tourist visits Jamaica not with the intention of staying at an all-inclusive north coast hotel or even a rustic little villa with family and friends. Typically he operates alone or with one or two colleague buyers. His stomping ground is much wider than the coastal ribbon, embracing territory ranging from the heart of urban centres to the most isolated of hamlets in the hilly interior. The objective is to ferret out vinyl and take or send it back home.

The modus operandi, for the more aggressive and brave among them, is to fan out across the countryside and "rough it," often accompanied by their local agents, knock on doors, and enquire of older folk, in particular, whether any of their vintage records are still around. Alternatively, the vinyl tourists may depend solely on local dealers to do their foraging and on their visits make contact with their agents.

Indications are they are mostly Japanese, German, French, Brazilian, and to a lesser extent British. The reason for less direct British involvement is possibly explained by reggae, broadly defined and its exponents, having been exported en masse to the United Kingdom in the sixties, seventies, and beyond. Considerable volumes of British-pressed-and-labelled Jamaican reggae records reside in the United Kingdom. In a sense, then, the British could stay at home while "raiding" Jamaica's vinyl cupboards.

It seems fair to say the vinyl tourist is a low-budget visitor for everything *apart* from purchasing old records. In other words, he is not here for relaxation or fun in the sun; he scrimps and saves on lodgings, transportation, food, and entertainment just to ensure that as much of his money as possible goes to the acquisition of vinyl records. In effect he is a long-stay business tourist with a decidedly nonexecutive profile, but, as will be borne out in the study, his financial resources can be considerable and easily outstrip those of the "traditional" tourist.

The Support Infrastructure

In an informal but very effective manner there exists an institutional arrangement that facilitates vinyl tourism. This is in the form of a cadre of local vintage-vinyl dealers; middlemen who function in an at once loosely coordinated yet closely networked relationship. Competition among the dealers is fierce, yet for self-serving reasons there is at times grudging cooperation within the ranks.

Given the nature of the business, it is impossible to get an accurate measurement of the number of dealers, but word from one of their number is that the complement is somewhere in the region of thirty. Some of the less energetic functionaries confine their activities to the main urban centre, Kingston, but it is generally appreciated that rural districts offer more for less risk in terms of safety and "ease of negotiation." Rural folk are still generally perceived as being gentler (and perhaps a bit more gullible).

As far as the author can ascertain, with two exceptions at most, the dealer "phalanx" is an all male cohort of largely working-class individuals. One of the two female dealers, both of whom effectively own their businesses, operates from the relative security of a store in suburban Kingston, or "uptown," as it is styled. This is not typical; the majority of dealers are itinerant in the sense that they don't operate from shops but rather from cars or occasionally from the facility of a "rival colleague" who runs a store.

The dealer network provides transportation, "cultural orientation," and "legitimacy"/protection for vinyl tourists as they make their way around city and countryside in search of their quarry. Alternatively, they take orders from the less adventurous tourist who may prefer to deal directly only with their regular agents. Beyond that, as symbiotic dealer/tourist relationships grow, some dealers provide lodgings for vinyl tourists.

Transactions are highly informal and little documented. When goods and money change hands, no written receipts are tendered, but the typical dealer's memory for detail and facility with arithmetic are quite impressive. By and large, dealer knowledge of Jamaican records is reasonably sound. Still, they are not given to thorough research and do not avail themselves of the gold mine of market information available on the Internet. Hence, occasionally they pay the price for ignorance with little-known high-valued sides slipping though the cracks and earning for them only meagre returns.

Indications are that local dealer knowledge of sought-after foreign idioms such as northern soul, jazz, and rhythm and blues leaves much to be desired. They tend to combine all foreign music styles into a homogenous mass, which is referred to in Jamaican vernacular as "souls." This places the local dealer network at the mercy of shrewd buyers, and the Japanese in particular are noted for their savoir-faire and proclivity to take advantage of knowledge-deficient sellers, whether dealers in their capacity as middlemen or the record owners themselves.

The Product Range

Unlike typical First World citizens, the overseas aficionados of vintage Jamaican music, a sophisticated and savvy cohort by all indications, and the local dealers that supply them, are acutely aware that there is so much more to reggae than Bob Marley. Despite his international stature and the reverence accorded him, Bob Marley is but one star in the reggae firmament. It has been claimed repeatedly (and quite likely never refuted) that Jamaica has produced more records per capita than any other country.

Certainly during the nineteen sixties, when the economy was relatively buoyant and the range of entertainment narrower (radio and dances were premier forms of recreation), if the occasional sales figures released by record producers were to be believed, record sales were at *record* levels. Indications are, despite economic downturn, the seventies also yielded bumper crops of records (but apparently with lower average sales) with artistes enthusiastic to articulate heartfelt issues in the socially and politically turbulent milieu of that decade. Decline in numbers set in during the eighties in response to economic conditions and competition for vinyl in the form of cassettes and later compact discs. The proliferation of digital media has seen this trend intensify to the present.

The following is at best only a sketchy profile of the extensive body of Jamaican recorded music from the point of view of the various record producers, labels, singers, and musicians most sought after by the overseas vinyl-buying public. From a producer perspective, Jamaica enjoys a high profile in the world vintage vinyl market for the efforts of the following:

- Clement "Sir Coxsone" Dodd. whose Studio One outfit is regarded as Jamaica's all-time leading record production centre. One UK estimate based on a cataloguing exercise has it that Mr. Dodd produced some eight thousand 45-RPM singles and over five hundred long playing (LP) records
- Arthur "Duke" Reid, whose Treasure Isle stable competed hotly with that of Sir Coxsone for market leadership. Indications are that his discography does not approach that of his archrival but on a "pound-for-pound" basis is at worst not far behind
- Cecil "Prince Buster" Campbell, a prolific independent producer who audaciously challenged the two abovementioned stalwarts with almost mind-boggling fecundity, turning out records literally by the hundred, by one UK estimate about six hundred in eight years, an impressive count for an indie
- Aside from "the Big Three," a plethora of second-tier producers, including: Lee "Scratch" Perry, the Khouri family, Randy Chin, Clancy Eccles, Derrick Harriott, Sonia Pottinger (by far Jamaica's leading female record producer), Byron Lee, Edward "Bunny" Lee, Edward Seaga (a former prime minister of Jamaica), Vincent "King Edwards" Edwards, Blondel Keith Calneck (Ken Lack), and worthy of special mention, Justin "Phillip" Yap, whose Top Deck family of labels is perhaps, pound for pound, the most important and strongest of the essentially ska-focused imprints, and Leslie Kong, unique among leading Jamaican producers, as far as the author knows, for recording on a single imprint, the Beverley's label. He first recorded Bob Marley (as Bobby Martell!)
- By record label, the vintage vinyl seeker will go for, inter alia:
- The wide range of Coxsone imprints (some of which represent collaborative productions), including World Disc, All Stars, Port-o-Jam, Studio One, Coxsone, Coxson, Wincox, Supreme, ND, C&N, Rolando & Powie, D Darling, Ironside, Bongo Man, and Money Disc
- The Duke Reid stable imprints, Treasure Isle, Duke Reid, Duchess, and Soul Shot
- The array of Prince Buster labels including Prince Buster, Shack, Soulsville Centre, Islam, Olive Blossom, Buster Wildbells, and Voice of the People
- The small Top Deck family of labels, namely, Top Deck, Soundeck, Tuneico, and Top Hat
- The Randy's label, which turned out some of the most hard-to-find ska by leading lights on the sixties scene. Subsidiaries include Pat's and Chappy

- The Ken Lack Caltone label and its stable mate Jon Tom, which constitute one of the strongest rocksteady brands
- An array of other producers' imprints, including High Note/Gayfeet, SEP, Pussy Cat, Golden Arrow, Head of Gold, WIRL, Top Sound, Caribou, Federal, BMN, Gala, BRA, King Edwards, Wail 'n' Soul 'm,' Merritone, Anderson, Links. and Matador
- UK pressings of Jamaican records on labels including Blue Beat, Island, Black Swan, Doctor Bird, Mellodisc, Punch, Blue Cat, Duke, Nu Beat, Escort, and Bullet, available locally in appreciable quantities as the legacy of Jamaicans' sojourns in Britain
- "One-off" labels, often corporate-sponsored, for example, Bata, Liquid Foods, and Machado Tobacco Co.
- By singer/musician the market seeks after:
- The Skatalites, by name and nature the quintessential ska band, and their leading soloists Don Drummond (trombone) and saxophonists Roland Alphonso and Tommy McCook, on all the labels for which they recorded (i.e., most of the labels listed above)
- Other seminal ska bands, including the Granville Williams Orchestra, Carlos Malcolm and his Afro Jamaican Rhythms, and the Mighty Vikings
- Master solo instrumentals the likes of trumpeter Raymond Harper and guitarists Ernie Ranglin and Lyn Taitt
- The world famous Wailers whether led by Bob Marley, Junior Braithwaite, Peter Tosh, Constantine "Vision" Walker, or Bunny Livingstone
- Prince Buster in his singing/"preaching" style
- Among the relatively few females, Dawn Penn, Phyllis Dillon, Marcia Griffiths, Millicent (Patsy) Todd, and, to a lesser extent, Yvonne Harrison enjoy good market profile. Rita Marley and her group the Soulettes are also sought after, no doubt largely by association with the Bob Marley and the Wailers
- The ska and/or rocksteady solo luminaries such as Lord Creator, Derrick Morgan, Stranger Cole, Jackie Opel, Lee "Scratch" Perry, Ken Boothe, Alton Ellis, Larry Marshall, and Delroy Wilson
- The top tier of harmony groups the likes of the Paragons, the Melodians, the Gaylads, the Heptones, and the Maytals
- Other ska/rocksteady/revive-reggae singing aggregations such as the Kingstonians, the Ethiopians, the Pioneers, the Tennors, The Bleechers, the Mellotones, the Rulers, the Termites, and the Silvertones.

As noted before, the upper-layer artistes and producers do not exhaust the universe of interest and pursuit for the vinyl tourist. Little-known artistes on obscure labels can command handsome prices simply because the vinyl tourist is a creature seeking to cater to the dictates of an adventurous international buying public intent on research, knowledge expansion, and acquisition of that which it discovers. A reality of the music business is that many good productions fell by the wayside commercially due to lack of effective promotion; thus, copies exist in very small numbers. Some have been discovered and command respect and good prices among vintage vinyl collectors.

Outside the realm of reggae broadly defined, there is also what might be called a "microniche" market for the calypso and mento idioms which predated ska. These count among some of Jamaica's earliest productions and the seminal efforts are found largely on shellac 78-RPM records dating back to the late nineteen forties and produced by sound pioneers such as Stanley Motta and later the abovementioned Arthur "Duke" Reid.

An unlikely yet increasingly important element of the Jamaican vinyl tourism product is not Jamaican in origin at all. Jamaicans are historically world travellers, their sojourns having taking them mostly to the United States, the United Kingdom, and Canada. Jamaicans are also, by and large, music lovers, not least lovers of Black music idioms such as rhythm and blues, its more polished counterpart soul, including the highly sought-after northern soul, and jazz, which has been described as African American classical music.

Upon returning from either short visits or long migratory stints, Jamaicans have brought back with them precious stashes of original-label recordings in the abovementioned idioms, some on small independent labels that never saw the light of day at home through local pressing and distribution. These have recently found a market among vinyl tourists, so broadening the product base and prolonging the potential life span of vinyl tourism. Also, the distinct possibility exists that, with the destruction wrought on New Orleans in August 2005 by Hurricane Katrina, many prized and priceless records of the jazz, blues, and rhythm-and-blues genres were irretrievably lost, making Jamaica, in the eyes of some vinyl prospectors, a relatively more important hunting ground for such platters.

Product Distribution

While by land area, Jamaica is tiny in the global scheme of things, its social demography is not homogenous. Hence, the various genres of vintage music are not necessarily distributed evenly across the island. By and large what will be termed "hard-core" reggae (in its broad definition), the product of working-class producers and musicians, can be found mostly in the poorer sections of the urban centres and in small rural communities. In the more affluent sections of Kingston, a typical oldies collection will feature foreign records, calypso, and lighter reggae such as the ska of Bryon Lee, with its quicker tempo, pronounced top end, and less emphatic bass line; and some productions of the Khouri family, whose Federal Records stable was a leading source for "uptown" reggae, that is, the musical creations of relatively well-educated and light-skinned Jamaican artistes.

In the cool mountainous interior, in particular the parish of Manchester and its relatively wealthy capital Mandeville, there are many residents with "British interests," either returned residents or expatriates, some of whom happen to have acquired reggae-inclusive collections back home in the United Kingdom. There can be found UK pressings of Jamaican offerings typically in above-average condition consistent with the ethic of preservation these persons embrace. Unlike that for "three S" tourism, the vinyl tourism product is more widely distributed, with much greater concentration in the interior.

Elite interviewees' perspectives on vinyl tourism

As a preamble to this section, it must be stated that, consistent with Jamaican culture of according nicknames, all names used below (aside from "Randy's") are pseudonyms.

The dealer side

A Downtown Kingston Hub for Vinyl Tourists

During the nineteen sixties, those halcyon days when vinyl held sway as the primary medium for recorded popular music, Randy's Record Mart, situated at North Parade in the heart of downtown Kingston, was Jamaica's largest retail record outlet. The record shop was situated on the ground floor, and the famous Randy's Recording Studio graced the upper floor.

But with the decline of vinyl and waning fortunes for the Jamaican economy, that outfit has undergone metamorphosis courtesy of founder migration and a name change (although, in the minds and hearts of most, the old name won't die easily). Still, the little outfit is run by bona fide successors to the Randy's empire and its attendant goodwill and *aura;* persons who literally grew up with the business.

No surprise, then, that this locale is something of a hub for vinyl tourists. Its unpolished urban feel and its bona fides as a treasure trove of musical information impart to it a rough charm and appeal to those who wish to sample Jamaica in the raw, acquire a bit of its musical history in conversation and plastic, and not be titillated by the frills of an overbuilt luxury environment abutting a white-sand beach.

What better source, then, for authoritative insight into the roots and progression of vinyl tourism? Thus, two of the proprietors were interviewed. Understandably, since they largely work together as their stories run parallel and their responses can be treated almost as coming from a "composite individual." In effect, this sets the stage for "expert witness or informant" to the little-publicized vinyl-tourism phenomenon.

It must be appreciated that these veteran downtown retailers isolate themselves splendidly from the pack of itinerant dealers. Given factors of age, experience, a dignity averse to the intensely competitive approach of the "street dealers," and the privilege of a fixed place of business backed by proud history and its attendant status, they represent something of an aristocracy and are elder statesmen among record dealers. Their account follows.

THE GENESIS OF VINYL TOURISM

Indications from these veteran dealers are that the invasion of vinyl tourists began in earnest around 1977. What we call roots music was in its heyday then with a populist socialist administration in power and with lyrics highlighting the works and words of Black activist Marcus Garvey and the perceived deity Ras Tafari (Emperor Hailie Selassie of Ethiopia). Given the nature of the lyrics, pioneering vinyl tourists classified many of the songs, delivered by artistes such as Burning Spear, Culture, Earth and Stone, and the Mighty Diamonds, as "spirituals" or "revival."

All in all the timing does not beg for particularly deep analysis; it coincided with the ascendancy of Bob Marley as a reggae exponent of global stature as his "Rastaman Vibration" and other albums commanded international attention. Marley electrified the First World with his flashing dreadlocks and his audacious protest lyrics, albeit balanced with songs of "positive vibrations" and the age-old theme of romance. To the peoples of secular developed societies with their hard-nosed achievement-cum-acquisition-oriented worldview, Marley's aura was evidently one of *exotic spirituality.*

The Progression

Then, in the early eighties, the Japanese started arriving in earnest. At first they demanded mostly ska, then spread their wings and came forward in time to incorporate rocksteady and reggae "proper" in their searches.

In terms of their modus operandi, although they typically travelled together, that did not prevent Japanese vinyl tourists from jostling energetically among each other for the best records. As with the dealers, they appreciated the importance of "cooperation in competition." At buying sessions they would negotiate energetically and employ body language (such as winks and hand signals) to tip off sellers that they would top attractive and seemingly final offers from their rival companions.

The men from Japan would also prepare wants lists and leave them, along with mobilization money, for the dealers to go out in search of the desired discs. In terms of establishing their presence, they would on first visits book into large hotels, but after familiarising themselves with Jamaica and getting a feel of the culture, they would feel comfortable to lodge at smaller, less formal facilities. By way of getting around, they would hire taxis to traverse the island, seeking out sound-system operators and private householders.

According to the veteran dealers, the Japanese would often pay little relative to the growing market value of the records they identified (a few were even lucky enough to have gifts made of the records). They had the edge and they knew it; the typical elderly widowed housewife would not have a clue as to the value of old records bequeathed by her late husband. So usually she wouldn't haggle much if at all with the oriental business-men, for ultimately those space takers would simply be tossed out anyway. A few thousand Jamaican dollars for an aged carton box with two or three hundred dusty, fifteen- to twenty-year-old 45s seemed like a fair price.

The Japanese would come on average three times yearly, but not in December (which marks the start of Jamaica's "regular" winter tourist season) because at that time record sales and auctions were held in their home country.

Other nations represented prominently were the United Kingdom, Germany, France, Brazil, and Bermuda. The typical length of stay in those pioneering days was one to two weeks.

An interesting information titbit volunteered by these pioneering dealers has to do with a financially beneficial offshoot to Jamaican artistes of previous decades. Japanese vinyl tourists have been known to procure the services of popular 1970s artistes to perform at stage shows in Japan.

As fixed-abode agents who have paid their dues, the Randy's crew now appear to entertain a select clientele of vinyl tourists. They are no longer so much in the business of tramping the streets and beating the bushes for vinyl; typically, records must come to them.

An Aggressive Semi-Itinerant

Skorpion is an aggressive but personable vintage-vinyl dealer of twelve years' experience. He is blessed with boundless energy and a genuine generosity of spirit that makes him impossible to dislike despite his occasional manifestations of "urban smarts." He got into the business quite by accident in 1994, having acquired a little stack of old Jamaican records by the way. The records came to the attention of three Californians with whom he was working in his capacity as a recording artiste. They made a surprisingly handsome offer for the collection and asked for more, alerting him to the money-earning potential of oldies and whetting his appetite for making a go at vinyl dealership. Now he runs a small no-frills store in midtown Kingston, but this facility is by no means a nine-to-fiver. Understandably, he spends a great deal of time on the road and in the "bush," so his shop is largely a storage facility and a venue for prearranged visits by customers.

Although essentially a "sole-trader," Skorpion networks functionally with a significant percentage of the dealer cadre and usually goes record hunting with a partner or two. He also collaborates on an occasional or "one-off" basis with persons on the periphery of the trade for purposes of transportation and identification/acquisition of record collections. His stomping ground is just about anywhere in countryside or city, including some of the tougher neighbourhoods of Kingston City, where the author understands that many dealers are reluctant to put a foot. Indications are, though, that the Skorpion does not venture much into the opulent suburban foothills and hills of St. Andrew north of Kingston.

Skorpion does business with vinyl tourists from several countries, among which he cites Japan (his leading source of overseas business), the United Kingdom, Germany, France, and Brazil. He is uncertain of the number (it can be fluid), but he cited five or six as regulars. They come for all genres of Jamaican music, from the sixties to the mid-nineties, but one in particular is a specialist buyer concentrating only on the music of the nineteen eighties.

Overseas customers keep in touch regularly both by telephone and frequent purchasing trips. By his account, some of Skorpion's customers visit the island on average every two months, and others visit on a quarterly or semiannual basis. As cited in the subsection "Profile of the Vinyl Tourist," being here for business purposes of a peculiar nature, they are not short-stay visitors. The average length of stay is two weeks, but, depending on the success of their prospecting, they may prolong their visits to three weeks or even a month. He will on occasion accompany them on buying trips. Skorpion emphasized that his customers do not go to the beach.

Another service he provides is to box and ship records overseas for customers who make their selections while on their relatively extended stays. Skorpion's customers pay him either in U.S. dollars or Jamaica dollars. Now this hardworking dealer is on the receiving end of growing orders for foreign records in the jazz, blues, and rhythm-and-blues categories, a phenomenon that emerged about a year ago. Significantly, though, there is to his experience very little interest in Jamaican-pressed foreign records; his overseas buyers want original foreign presses only, the type cited earlier that returning Jamaican residents brought home.

Perspectives of a Suburban Kingston Dealer

Andy is a lanky thirty-something vintage vinyl agent whose boyish looks belie his years. He is something of a "social outlier" among dealers, not being from a working-class background but possessing solidly middle-class credentials and residing in a middle-class uptown neighbourhood. These credentials he uses to good effect for acquiring stock from certain communities. On venturing uptown, he sports conservative dress and his "appearance" (dress-cum-complexion-cum-speech) is something of a passport to the homes of the genteel elderly in pristine uptown neighbourhoods.

Andy, who has been in the business for a little over four years, nevertheless does not confine himself to suburban Kingston/St. Andrew but also prospects island-wide. In making frequent purchases uptown, Andy has proven his ability to locate hard-core reggae well north of Cross Roads (the area demarcating "downtown" from "uptown" in Kingston, the capital city), somewhat disabusing the author of an a priori notion to the effect that hard-core reggae was not to be found uptown.

Andy is quite familiar with the "Japanese connection," knowing of about six regulars and doing business with at least two on a reasonably frequent basis, including the eighties specialist identified by Skorpion. Andy offered a little insight into the purchasing power of some Japanese vinyl tourists. One confided to him that he had readily available for spending about three million Jamaica dollars. At the time, the Jamaica/U.S. dollar exchange rate was about sixty to one, translating into approximately fifty thousand U.S. dollars. This, he was told, could have been easily supplemented if buying opportunities exhausted it. It would simply be a matter of wiring for more. Andy also sells regularly to at least one Englishman.

His customers also pay either in American currency or with local dollars. Confirming the assurances of the other dealers, Andy indicates his customers are also not beachcombers. Substantiating the emerging trend cited by other dealers, Andy too notes the growing interest in foreign genres.

The Vinyl Tourist Side

A View from the French-Brazilian Connection

Frenchie is a thirty-something international citizen of sorts. This amiable soft-spoken vinyl tourist, as his moniker suggests, was born in France. He now makes his home in Brazil after having spent part of his childhood in Chicago, USA.

Frenchie made his debut on the Jamaican scene in January 1998. Since then, he has established a regular visiting pattern subject to the availability of funds. On one early trip, accompanied by a friend, he visited the beach. There has yet to be a second beach visit.

He recounts having flown into Montego Bay on his first trip, then making his way by bus to Negril. Negril did not offer what he sought, so he again boarded a bus, this time headed for Kingston. He ended up downtown, roaming about on his own, and stumbled upon Randy's at North Parade. That was the start of an enduring mutually beneficial relationship between him and the Randy's crew. They are one of his primary sources for Jamaican oldies on wax, and when in Jamaica Frenchie actually rooms with a member of the Randy's outfit. Interestingly, from the outset he never stayed in large hotels but displayed a preference for guesthouses or private homes.

On average, Frenchie visits Jamaica every three to six months, and his typical length of stay is three to five days, relatively short by comparison with Japanese vinyl tourists. He recalls his single longest stay was ten days, sometime in 2001.

Frenchie was not particularly forthcoming about the business side of his activities or any acquaintance with other vinyl seekers and appears far more interested in building a personal collection of reggae oldies, mostly ska, rocksteady, and revive reggae, from the source country.

A Perspective from the Japanese Connection

Hitachi is part of the "Japanese connection," the single most influential national presence on the Jamaican vintage vinyl scene. He is a diminutive, cheerful vinyl tourist who hails from rural Japan and made his debut visit to Jamaica in 1993. Driven by an early interest in Jamaican music, the primary purpose for Hitachi's maiden stopover was to "check out" local sound systems, but he laughingly admitted, between puffs on his cigarette and sips of his Red Stripe beer, to having visited the beach once on that earliest trip. As with Frenchie, there has been no repeat beach visit!

Having caught a glimpse into the business potential for trading in vintage vinyl and the opportunity for accumulating a personal stash of Jamaican oldies, this frequent-repeat vinyl prospector makes his way to Jamaica on average twice a year, and in addition his stays are lengthy, lasting up to three months at a time. Obviously that translates into Hitachi spending a fair amount of his time in the island. In fact, he admitted to one extended stay between 2002 and 2003. This savvy young businessman, realizing the importance of effective communication in his trade, signed up for English-language classes in Jamaica during this period of unofficial residency.

Not surprisingly, on his first few visits, he stayed at larger established hotels in Kingston, Montego Bay, and Negril. Then, having established his contacts and familiarised himself with the "runnings," Hitachi was soon able to make firm arrangements for home stays, and that is what invariably obtains now.

During his visits to the island, Hitachi arranges for his acquisitions to be shipped home. He purchases all genres of Jamaican reggae, from ska to dance hall. In respect of his private collection, he does not favour dance hall; ska, rocksteady, and early (revive) reggae are his preferences. On a personal basis, he also collects shellac 78s with pre-ska mento and calypso from the late nineteen forties through the early sixties.

Hitachi has never hung out a shingle in his home country but plies his trade over the Internet. This energetic young businessman is personally familiar with five or six fellow Japanese vinyl tourists. Some of these compatriots, unlike Hitachi, do not run only cyber stores but operate from physical facilities in Tokyo and other big cities. Indications are Hitachi does not go on hunting/buying trips with his fellow Japanese but depends on his network of local dealers to get him around and/or deliver goods to his doorstep when he is in his second home country.

In volunteering some analysis of the Jamaican vintage-vinyl market, Hitachi cited factors such as upward pressure on prices due in part to dwindling supplies of the choicest items. But he noted growing Japanese interest in foreign wax in the jazz, rhythm-and-blues, and soul fields.

Emerging trends and future prospects

An Exhaustible Product

Collectors go in a big way for original pressings on original labels, or at least early re-presses. That's where the bragging rights and by extension the money values reside. Obviously there can never be another first or second pressing of a forty-year-old ska or rocksteady recording, so, as selling and purchasing go on apace, the numbers will only decline.

For this reason, the vinyl tourism product is truly nonrenewable, and, given the undocumented nature of the product base, it is simply impossible to estimate the remaining life span of the "core" product. There are, however, three factors that could delay the ultimate demise of the vintage-vinyl market.

The factor already identified is the emergence of demand for foreign genres. Again, it is impossible to estimate the "population" of such records in Jamaica, but as buyer interest grows, demand in this area could also take some "pressure" off the Jamaican records. On average, records in the genres of blues, rhythm and blues, jazz, and (northern) soul enjoy much higher unit prices than Jamaican genres.

The second factor has to do with the possibility of records "at the margin" assuming pride of place as stocks of current premium products diminish. Therefore, second or third pressings or originals in less than pristine condition will possibly eventually assume the status of the "sought-after."

Third, despite the irreplaceable nature of the dwindling original stock, there is an element of replacement as the Japanese in particular increasingly embrace reggae/dance hall in all its forms. As cited earlier, indications are that the Japanese are now willing to pay top dollar for dancehall records of the nineteen eighties and early to mid-nineteen nineties; these too were produced in substantial volume and should take some appreciable time to exhaust. With the passage of time, later output will likely take on vintage status.

Market Forces

There are clear indications of upward pressure on vintage-vinyl prices. This is not only due to exertion from the demand side, with buyers bidding up prices. In addition to that, there are real impulses from the supply side. One is of course growing scarcity of product. The other is widening market knowledge. The secret that old records are at least reasonably valuable is being let out as dealers search further and wider for oldies. Now record owners are more aware, and the author knows of several instances where adult offspring have urged elderly parents not to part with their old records for a song (no pun intended). This upward pressure on prices will likely lead to greater revenue for owners/middlemen, for is not foreseen that vinyl-hungry foreign collectors will curb their appetite for acquisition, much less desist from their pastime-cum-source-of-income.

Economics Versus Ethics

Cultural Plunder?

The economics of the vinyl-tourism market is easy to understand. From both the owner and dealer perspectives there are returns to be made. For the record owner, there is a quick windfall (or occasional income if the collection is sufficiently large and suitable). For the dealer, the trade offers an income sustainable over a considerable period. Of course, there are positive indirect and multiplier effects; to taxmen and guesthouse operators, to the families of owners and dealers. It is just the free market at work. There are two ethical issues though. At the microlevel there is the question of whether owners/dealers are enjoying fair monetary return. Knowledge is truly power in this market, and it is the foreigners who, for the most part, have the knowledge edge. On the supply side, there are two levels of ignorance: owner ignorance and dealer ignorance. Compound them and the consumer surplus to the foreign purchaser can be astronomical.

At the macrolevel is the issue of an important and irreplaceable dimension of Jamaica's cultural heritage being stripped. This is roughly analogous to arguments about environmental degradation being visited on tourism hard-use areas. Here, though, we are talking about the export of tangible items that represent the priceless intangible of cultural history and lineage. The short-term gains are there for a few, but in the long run is it worth the price for Jamaica's best recorded works to reside elsewhere?

PARTING THOUGHTS

Vinyl tourism is a "dual-export" product involving at the same time the so-called invisible export of tourism per se and the tangible export of the finished vinyl products the tourists covet. It is real, but based on the nature of the activity, it seems doubtful it will ever be quantified with anything more than the roughest "guesstimates" being arrived at for its hard-currency contribution to the economy or share of total tourism earnings.

Based on information acquired from the Jamaica Tourist Board (Jamaica Annual Travel Statistics 2005) and the Economic and Social Survey of Jamaica (Economic and Social Survey of Jamaica [ESSJ] 2005) report of the Planning Institute of Jamaica, for 2005, average length of stay for stopover visitors was 9.8 days. Mean expenditure was US$1,014.34. For cruise-ship visitors, average expenditure was US$85.21. As cited above, from dealer/tourist accounts, there is evidence that the typical vinyl tourist's stay and spending far exceed the traditional tourists' averages.

Understandably, one cannot depend on either the local service providers or their overseas clients to be overly forthcoming about the nature/magnitude of monetary transactions involved. It lies squarely within the territory of the informal economy.

Again, let it be emphasized, the business is conducted discreetly; the numbers are impossible to pin down because neither buyer nor seller wants them revealed in their entirety, but undeniably the evidence for vinyl tourism as an established tourism niche within a niche is well beyond the anecdotal.

REFERENCES

Economic and Social Survey of Jamaica (ESSJ). 2005. Kingston: Planning Institute of Jamaica.
Government of Jamaica 10 Year Tourism Master Plan. 2001. Kingston: Ministry of Tourism.
Jamaica Annual Travel Statistics. 2005. Kingston: Jamaica Tourist Board: Corporate Planning and Research Department.

5 Tourist Nationalism in Trinidad and Tobago

Raymond Ramcharitar

INTRODUCTION

The British novelist Arthur Calder-Marshall, visiting Trinidad in 1938, observed the curious postures of hosts and the metropolitan tourist:

> The desires with which he may have started, of getting to know and understand the life of the Negroes, quickly disappear . . . He spends his time with people of his colour. . . . He learns that it is not done to ride in buses or drink in the rumshops. The Negroes he meets are servants, whose natural independence has been corrupted by low wages and big tips. He dislikes them and generalizes his dislike to include all Negroes and East Indians in the island (Calder-Marshall 1939: 94–96).

The implication here is that to the group at the top of the local hierarchy, in this instance the small White economic elite, descendants of British and European migrants, tourism was merely one form of skirmish in the ongoing battle for cultural hegemony. To manage the tourists' experiences, to convince them, and the outside world, that the fundamental nature of the place corresponded to their idea of it, was a victory which endorsed the local rulers' hegemony and furnished a narrative for export.

This desire, and the way of realizing it, issued from the dynamics of a heterogeneous society of many ethnic groups, with competing values and cultures, and the need for the small White minority to assert their cultural hegemony. Today in Trinidad, the circumstances have changed: the foreign visitor can no longer avoid local colour, and is not encouraged to, but the politics of perception remains the same. Though the status of the competing groups has shifted, the battle being fought, sometimes under the misleading rubric of "tourist promotion," has not altered: the promotion of a particular type of knowledge about Trinidad, a desire that Trinidad be known in a particular way, ostensibly to validate the "Creole"[1] postindependence national culture.

A good example of a repetition of the dynamic described above, in the service of the new exigencies, was available in 2003, when the U.S. E! Television produced an instalment of its "Wild On" Series on Trinidad Carnival. The programme aired on E! Television on June 28 and precipitated outrage among Carnival aficionados in letters to the editor in the island's three daily papers and call-in radio talk shows. The *Trinidad Newsday*'s editorial summed up the general feeling: "We are compelled to express our outrage and to demand that TIDCO [the Tourism Authority] seek from E! TV an apology for such an insult to our country and its culture" (*Trinidad Newsday* June 27, 2003, 10)

The insult was embedded in the too many "White" and "light-skinned" bodies, and the events shown—parties and after-party yachting to the small islands off the north coast—were not recognized as "authentic." The ire prompted the government to commission a documentary from American Black Entertainment Television (BET), which, directed by state personnel, soothed local passions.[2]

This outburst was merely the latest in a pattern of such reactions to "unsuitable" representations of the country's culture, and particularly Carnival. Dutch anthropologist Peter van Koningsbruggen, investigating Carnival in the late 1980s, recounts several conflicts with Carnival officials who insisted on controlling the representation of academic and journalistic accounts and an attitude that was impossible to satisfy: "This tendency to reject almost every presentation or interpretation of the actual situation of Trinidad is translated into feelings of bitter indignation and offence which are always directed against the documentary makers" (van Koningsbruggen 1997: 212).

These illustrations suggest that the conflict issuing from dissatisfaction with the mode of representation in tourist discourses in Trinidad has little to do with tourism as it is understood conventionally. What is promoted as a discourse of commerce and development is a projection of another, more violent battle for cultural and political hegemony: the label of "tourist promotion" provides a mask and symbolic stage for this battle to continue. This process can be formally identified as a conflict of knowledges and dialectic as a means for producing a particular type of historical knowledge articulated in a specific way.

Conventional tourist discourse is disseminated via conventional mass media: advertisements in broadcast and print media in target markets along with the cultivation of exposure via travel writing, filmmaking, and popular media "location" events—like the Travel Channel, Discovery Channel, MTV, and the Internet.[3] But Trinidadian tourist knowledge differentiates itself from conventional tourist knowledge in two ways: first, it aspires to a more ambitious status and mode of exposure than mere commercialism. Rather than a populist festival, simulacra of which can be found elsewhere (like Brazil's Carnival in Rio), Trinidad sells its major tourist attraction, Carnival, as intrinsically historical knowledge, embodied in the authentic original of a cultural phenomenon which has spawned many satellite festivals in North America and Europe. Second, looking at local advertisements, it seems evident that Trinidadians are as much the targets of the narratives as foreigners. In describing the launch of a "tourism park" in Trinidad, a newspaper reported that "Produced by the Ministry of Tourism and Tourism Development Company (TDC), the Park is staged with the aim of creating impact and awareness of T&T's tourism product, in a manner that will capture the attention and engage the passion of our people" (*Trinidad Guardian* August 24, 2006: 38). The following month, advertisements appeared in the local press stating that locals could be considered tourists once they had the correct attitude.

The fusion of cultural nationalism and tourism is not unique. Historian Barry Higman recognizes that, in the English-speaking Caribbean, historical discourse has been fed into the popular or populist spheres via what he calls "heritage tourism," which precipitates the creation of tendentious historical narratives focusing on a particular event (like slavery, or maroonage in Jamaica). These narratives and the media developed for their dissemination (music videos, posters, backdrops in film locations) are easily digestible and saleable and have the potential to transform the region into "a kind of wall-to-wall museum piece, to be viewed through the framing glass of the tourist's air-conditioned bus window" (Higman 1998: 249). Though in Trinidad's case, this framing is as much for the native as the outsider.

The production of a glib populist history in the service of tourism has been a part of the Trinidadian political culture since independence. Derek Walcott, who lived in Trinidad from the late 1950s to 1981, observed that the "steelbandsman, carnival masker, calypsonian, limbo dancer" were "trapped in the state's concept of the folk form . . . the folk arts have become . . . an adjunct to tourism since the state is impatient with anything it cannot trade" (Walcott 1998: 7–8).

In the period Walcott describes, there was also the deliberate attempt to fuse the emergent national consciousness with a particular set of symbols: Carnival, the steelband, calypso and the "folk" were associated with Afro-Trinidadians, and aligned with a particular political party, the Peoples National Movement (PNM), from the moment of independence. The party's uninterrupted 30-year stint in government (1956–1986) allowed it to entrench its worldview in various institutions. As anthropologist Kevin Yelvington notes, "Eric Williams and the PNM took up the banner of 'creolisation' but this ideology constructed 'Trinidadian' and 'national' practices as Afro-Trinidadian 'culture' and labelled practices (such as 'East Indian' culture) which deviated from such a process, as 'racist' and 'unpatriotic' " (Yelvington 1993: 13).

It is useful to point out that members of a particular social-ethnic group rooted in this party—the "brown" bourgeoisie of Trinidad—were the architects of this discourse. As Peter van Koningsbruggen, in his study of Carnival, notes, "Creoles" were distinct from the black underclass and working-class masses, and their political "dominance was largely based on the tactical manipulation of the traditional racial and social rivalries between blacks and Indians, accentuated by the fact that the [political] opposition remained Indian in composition" (van Koningsbruggen 1997, 119; also Balliger 2005a, b).

Having briefly discussed the links between tourism, cultural nationalism, and politics in Trinidad, this chapter will attempt to:

- examine the "tourist" discourse generated by recent events, including the promotion of the Trinidad and Tobago presence in the 2006 Football World Cup Finals in Germany;
- discuss the conflicts tourist discourse and narratives mask, suppress and omit;
- examine the production of tourist and academic discourse relating to Trinidad Carnival; and
- identify the social group which has engineered the cultural nationalist discourse from which tourist discourse issues and outline its current status and projects.

CARNIVAL, TOURISM, AND NATIONALISM IN TRINIDAD

As indicated in several other chapters in this volume, tourism is crucial to Caribbean economies. As former Jamaican Prime Minister Michael Manley noted: "The governments and leaders of the Caribbean have finally and almost unanimously come to the view that tourism is anything from 'an important' to 'the most important' to 'the only' means of economic survival for their states" (Patullo 1996: ix). But Trinidad is the exception: it has no real need for tourism since its oil and natural gas have provided, for a century, a reliable and steadily increasing source of revenue. Furthermore, with the boom in world crude oil and gas prices post-2001, money has not been a problem in Trinidad.

In the words of IMF auditors, the country "is one of the larger Caribbean nations in terms of GDP and population. Its energy wealth has enabled it to achieve one of the highest standards of living in the region. Nevertheless income and inequality and poverty rates are only at about regional averages" (IMFa 2006). The per capita GNP in 2004 was US$9,479, almost on parity with Barbados (US$10,276) and well above the other significant economies—Jamaica (US$3,013) and that of the combined Eastern Caribbean Currency Union (US$5,757) (IMFb 2006, 5). Trinidad's GDP growth since 2001 has been supernormal: the GDP in 2000 was US$8,153M; in 2003, it was US$ 10,502M (Central Statistical Office)[4]. And yet, five years after the post-2001 oil-price spike, which coincided with the accession of the PNM government after a six-year hiatus in opposition, a national newspaper commented:

> The murder rate has increased dramatically over the past five years—
> in fact, there has been a 300 percent rise in killings. Nor has the rate

in other crimes shown any decline in recent times. Between 2004 and 2005, woundings and shootings went up by 30 percent; rapes and other sex offences by the same; burglaries by 15 percent; robberies by nine percent; and white-collar crime by 14 percent. But even these increases must be interpreted in light of the fact that, when violent crime is prevalent in a society, other crimes tend to be under-reported (*Trinidad Newsday* November 16, 2006, 10).

A newspaper editorialized about the public transport which, without illegal taxis, "the system may well collapse," even as it mentioned that sometimes "criminals pose as legitimate taxi drivers only to turn around and rob and rape passengers" (*Trinidad Express* August 17, 2006: 10). In late 2006, the government announced plans to import food because of spiraling inflation and underproduction—inflation was at 10 per cent, and food prices had risen by 26 per cent that year (*Trinidad Express* November 24, 2006: 1).

In the wake of the 2006 Football World Cup campaign, yet another newspaper noted that while German interest in Trinidad had "grown exponentially," there were many "challenges," in the country's shortage of hotel rooms, its decaying infrastructure (especially transportation), "security" and "quality of service" (*Trinidad Guardian* June 15, 2006: 28). Additionally, related tourist services seem to be deficient across the country.

From all this, the *economic wisdom* of encouraging an increasing stream of visitors to such an environment—where transportation, food, and public security systems are dysfunctional and money is plentiful from other sources—to promote and commodify local culture is dubious, if normal developmental goals are the objective. There is evidence, though, that normal developmental goals, for which tourism is used as a signifier, are not the objective of tourist nationalism. This evidence is to be found in the historical strategies of the ruling party to retain power and its specific posture toward tourism.

The PNM took office in late 1956 and formed its Division of Culture in 1957. The role of a cultural programme in its governing strategy has been noted by various scholars. Gordon Lewis noted "the official recognition of Carnival by the Peoples' [sic] National Movement government as a national heritage," whereupon "the calypsonian, like the steel band leader, becomes a symbol of the new national pride. A new state cultural apparatus appears, to organize cultural standards. He is told he is important to the tourist industry . . ." (Lewis 2004: 23).

Gordon Rohlehr describes the *political management of culture* by the PNM, how its apparatus "inscribed in the national mind the notion that cultural production has, like everything else, become one feature of an enormous system of spoils." This was enacted through programmes like the "Prime Minister's Best Village Trophy Competition" in which rural villages competed in areas like handicraft, concerts, and sanitation. However, researchers observed that "many people consider[ed] it as little more than a tool by which the ruling party has established a tight control over villages" (Rohlehr 2004:143–5).

Assessing Carnival and society between 1970 and 1984, Rohlehr noted that the culture had the effect of the "equating of intellectuality with impotence" and precipitated the rise in genres of calypso like the "patriotic" calypso, whose purpose was to "preserve the sheen on the charisma of the maximum leader in an atmosphere of collapse and disorganization at almost every level of social life" (Rohlehr 1992:326–7).

All this was reinforced by the already discussed idea of ethnic primacy and the culture was encoded, as suggested above, in Carnival culture—a system of symbolism and practices which codified the PNM's desire to retain power.

With the loss of power in 1995,[5] and its reacquisition in 2001, this cultural political system was reactivated. According to anthropologist Robin Balliger, the Indo-party's (the United National Congress [UNC]) accession "challenged the common perception of the Caribbean as the domain of the African diaspora," and the new government and the new visibility of Indoculture "were read by many AfroTrinidadians as an assault on Creole hegemony" (Balliger 2005a:184, 195). This generated a reevaluation of cultural perceptions to conclude that the independence dispensation had characterized "a national emphasis on the 'black man' and the 'black man's tings [sic]' ," which created a discourse of "hypernationalism," based on ethnic primacy of things AfroTrinidadian (Balliger 2005b:226)—this meant Carnival and satellite activities which fell under the catch-all of "local culture."

Upon the return to power of the PNM, opportunities arose to fuse this conception of national culture with an imperative of material national importance: tourism and thus economic survival. The discourse so produced could thus be disseminated with the imprimaturs of the state, business, and civil society. The media responded by intensifying their foregrounding of images and coverage of Afrocreole culture, like Carnival, Caribbean music concerts, personalities, and promoting a nationalism which was implicitly linked to the Afronationalist worldview (Ramcharitar, 2005). Indeed, post-2002 (and from a few years before), various television and radio stations emerged that explicitly promoted this cultural agenda: Synergy TV, Tempo TV, NCC TV, Power 102 FM, and I95.5 FM. A few "Indian" radio stations also emerged which played Indian music and foregrounded cultural events, like 103 FM and 90.5 FM.

In view of all this, it is instructive to note the attitude of press and populace to the UNC's staging of the (May) 1999 Miss Universe pageant, which included an international television link during the pageant via which the country's attractions would be broadcast to potential tourists. There were many criticisms, dismissal of the tourist exposure, and suggestions, via press and radio, as to what could be done with the money spent on the pageant.

Trinidad and Tobago's qualification for the 2006 Football World Cup Finals in Germany, however, was different, and provides an opportunity to examine the reification of this post-2001cultural codification. The news was met by the government with a media campaign and a determination that Trinidad's presence would provide an opportunity to create a "brand" Trinidad for exposure during the team's games and a symbol of "nationalism" at home. Constituent parts of the branding exercise were to include "the finest steelband and Soca acts to put on attractions before the game days in the three German cities to sensitise [sic] the people there" (*Trinidad Newsday* January 20, 2006: 5).

A government-sponsored newspaper pullout revealed that the government had commissioned a video, and jingle, and sent a contingent of some 128 people to the event, comprising Carnival bands, calypsonians, steel-pan sides, and a number of performers (*Trinidad Express,* Section 4, June 18, 2006). And the promotions were replicated at home: the sports ministry (to give a single example) announced that it had spent $2m in "giveaways ahead of the World Cup." The ministry's permanent secretary was reported as saying "the effort was an important one which allowed citizens to determine for themselves how they would display national pride for the Soca Warriors" (*Trinidad Guardian* June 17, 2006: 7.) One academic articulated the sentiments promoted in many ads and by government apologists: the campaign had precipitated a successful "sports nationalism," which she endorsed (*Sunday Express* June 18, 2006: 11). Other apologists, academics, and public figures and companies, via advertisements, echoed this.

Whether the intention of these activities was to promote the country's economic affairs is not certain. A newspaper editorial asked, after the government announced its intentions to send a trade mission to Germany after the World Cup, why this strategy had only just occurred to them: "Could no one in the government have imagined it before now, thought about the mission's design and objectives . . . and fleshed out an action plan with budgets and time plans?" (*Trinidad Guardian* June 18, 2006).

The 2006 Football World Cup Finals was a unique event, but when Trinidad Carnival cannot go to the world, the world is brought to it: Carnival 2006 saw a number of international media invited to Trinidad to cover the festival and a high-visibility media campaign to apprise the local public of this fact. In addition, there were (and continue to be) international steel-drum events designed to bring attention to the Trinidadian "national instrument," and the hosting of various academic and quasi-academic conferences on Carnival and the steel drum. One of the most ambitious such activities was the second World Steelband Music festival which was held at New York's Madison Square Garden in June 2005. Ten months later, it was reported that no prize money had been paid to winners (*Trinidad Express* April 22, 2006: 7).

The money mismanagement in Carnival business is legendary: the story is told in headlines like "Pan Trinbago seeks $100,000 Bailout" (*Trinidad Guardian* February 18, 2005: 6); "Despite Promises, Pan Body Homeless" (*Trinidad Guardian* February 26, 2005, editorial); "Love Affair Gone Sour: Steelbands

Struggle to Survive" (*Trinidad Guardian* February 15, 2006: 12–13); "How Pan Trinbago Fell Short" (*Trinidad Express* February 19, 2006: 9).

The disorganization is not restricted to Carnival but pervades state-sponsored cultural activities like Carifesta (Caribbean Festival of the Arts), which was staged in Trinidad in September 2006. The postmortem *Express* editorial was headlined "Caught with Our Pants Down," citing "complaints from overseas participants concerning travel and accommodations, and the low levels of attendance" in a festival that was in "partial collapse" during its execution (*Trinidad Express* October 1, 2006).

Yet these events persist and thrive. Two obvious reasons seem evident: the first, from what has been said, is that the political value of these cultural events is foremost, as opposed to their smooth execution—and political needs usually overwhelm efficiency. The second reason, in the view of this author, is that the events are exercises in mass distraction, to divert the population, and importantly, the media's attention from the realities of the national situation. This was made obvious in the Soca Warriors campaign: there was a months-long advertising campaign which disseminated the images, symbols, and positive appraisals of Trinidadian national culture locally. Indeed, the coverage of Carnival was replete with "positive" stories of parties and events. It is worthy of note here that in all the promotional materials—the advertisements, documentaries, commentaries, even the branding ("Soca Warriors")—the absence of the non-Afro-descended part of the population was noticeable. When a member of a local IndoTrinidadian organization suggested that perhaps the national football team be rechristened the "Chutney-Soca Warriors" so as to recognize the Indo-element of national culture, this was labelled "racist" (*Trinidad Express* January 11, 2006: 11).

The effects of this torrent of cultural national discourse, despite the failure of the actual events, led to public sentiments like "The evidence that arts and culture is [sic] our métier is overwhelming." And from this premise "it follows as surely as day follows night that the population's predilection towards arts and culture in its broadest sense should be central and not peripheral to the national development process" (*Trinidad Express* editorial, February 11, 2005).

This reiterates the connection between culture, nationalism, and the welfare of the state. Taking this as a lead, the organization for pan affairs, Pan Trinbago, could now justify its existence stating that "We work for the state, we contribute to the overall windfall of approximately six hundred million dollars, which is collected by the Treasury as taxes—this figure was quoted by the Minister of Tourism . . ." (*Trinidad Newsday* April 11, 2006: 11). This new "tourist" logic might also explain why a (Hindu) Divali magazine, in 2006, would be introduced by an editorial titled "Temples and Tourism in Trinidad and Tobago."[6]

The preceding gives some indication of the constituents of the tourist discourse, but it is questionable whether material returns vindicate the outlay in nationalistic sentiment and resources for promotion. Proof that the economic dimension of tourism, and its material elements (infrastructure and facilities), are not as important as the actual mechanics of tourist promotion on the national agenda could be found in the state of tourism facilities and the conduct of tourist events such as Carnival.

Magnus Linklater of the *Times* of London, in a letter to a Trinidad daily, complained about the "completely derelict and frankly dangerous state of the amenities at [a] star tourist attraction," and about the island generally, that visitors expect "well-maintained buildings" and "a well-maintained environment. Until Tobago is able to offer these [its] advertising slogan, 'clean, green, safe and serene' will be judged simply untrue" (*Trinidad Express* April 23, 2006: 13). The *Express*'s business section had announced some weeks before that "crime takes a toll on Tobago tourism" (*Trinidad Express* February 22, 2006: 16).

In Trinidad, Carnival now requires thousands of policemen and soldiers on the streets to keep it safe (*Trinidad Newsday* February 24, 2006: 3). During the 2006 Carnival season, police had to fire rubber bullets into a rioting crowd at a party (*Trinidad Guardian* January 30, 2006: 3). Apart from the disparity between the efforts at promotion and the unwholesome state of the objects promoted, other crucial questions remain of the efficacy of these promotional efforts, and government attitude and policy beyond the organization of seemingly spontaneous campaigns.

A 2005 report from the World Travel and Tourism Council (WTTC), produced in conjunction with the government and various state agencies, notes that "over the past 15 years (1990–2005), Trinidad and Tobago's Travel & Tourism activity has enjoyed a period of strong growth (from 1995–2000) followed by a more modest performance (since 2000)" (WTTC Report on Trinidad and Tobago, 2005: 24). This suggests that despite the intense post-2001 campaigns, tourism growth was stronger before 2001.

The use of BET to produce and market Carnival programmes resulted in the presence in Trinidad, in 2005 and 2006, of several African-American celebrities to sample the festival—these including movie stars Gabrielle Union, Anthony Anderson, Chris Tucker, and others, and at least one major musical collaboration, between Trinidadian soca star Machel Montano and international pop star, Wyclef Jean. The potential benefits of these intersections are considerable: the African-American travel market is huge and untapped, and other locations outside the Caribbean have been successfully marketed to the "Black" market (Patullo 1996:149–50).

But whether the presence of these celebrities, the concert in Madison Square Garden, and the sending of cultural contingents to Germany to "sensitise" foreign publics were accompanied by or would generate numbers of tourists cannot be determined by the data presently available (in 2006). Other data, however, allow assessment of the period 2002–2004, like the WTTC's yearly League Table on countries' expenditures and revenues from tourism and its report on Trinidad and Tobago.

The WTTC's report is optimistic on the country's tourism *potential*, but notes that "there is a long way to go to ensure that current negative perceptions of Travel & Tourism [by government and society] as a low-yield high-cost industry in both economic and social terms are overcome" (WTTC Report on Trinidad and Tobago, 2005: 7). It also states that as a result of the historical disregard, "Travel & Tourism development in Trinidad and Tobago has therefore been largely unregulated, unplanned and unsupported and, since the early 1990s, it has been driven more by regional and international trends and demand than by national policy" (WTTC Report on Trinidad and Tobago, 2005, 13). These external factors include terrorism, international currency fluctuations, high commodity prices which generate business travellers, and Trinidad and Tobago's insulation from hurricanes, which devastate other destinations.

The 2006 WTTC League Tables show that Trinidad and Tobago's 2005 expenditure on tourism amounted to 5.1 percent of its GDP, while Antigua and St. Lucia spent at least 17 percent of their GDP, Jamaica 16.7 percent, and Barbados 15.7 percent. Guyana, a far poorer country, spends 5.9 percent of its GDP on tourism.

This suggests that the government's apparent eagerness in promoting what are ostensibly tourist-related cultural events is not consistent with their overall focus, or lack of focus, on tourism—a deficiency which is reiterated by the government's development plans. In 2006, the government of Trinidad and Tobago was insistent upon setting up three aluminium smelters, which environmentalists and communities were resisting. Apart from the potential environmental devastation, a retired university academic, Prof. John Spence, pointed out alternatives, one of which was "tourism, based on the development of a nature park and to include all the other attributes of a beautiful, relatively unspoiled part of Trinidad. . . ." However, this seemed unlikely since "some of our political leaders . . . seem to believe that the only form of development is industrialization" (*Trinidad Express* November 23, 2006: 11).

As to the economic benefits of tourism, University of the West Indies (UWI) academic Keith Nurse (using 1998 data) reports a benefit-to-cost ratio for Trinidad Carnival of 7:1 (Ho and Nurse 2005, 329). In 2006, he authored a report that concluded Carnival generates US$126M (*Trinidad Newsday* November 20, 2006: 17). He does not reveal how these numbers are calculated, but there are indications that tourism might have hidden social costs which are not factored into cost-benefit analyses.[7] In 2003, the Trinidadian minister of tourism observed that the incidence of HIV/AIDS was becoming a serious and out-of-control problem in the islands since "sex tourism is growing and the 'beach bum' phenomenon in men and women is escalating in Trinidad & Tobago (T&T)." This was directly related to the tourism trade, said the minister, since "the quality of our tourism product is ultimately affected, as the health of the tourist worker is a major contributor" (*Trinidad Newsday* February 6, 2003: 3). The WTTC report also notes that "the potential economic impact of HIV/AIDS, which includes lost savings, investment, labour supply, employment and an increase in expenditure on health care, could amount to 5 per cent of Trinidad and Tobago's GDP by 2007" (WTTC Report on Trinidad and Tobago, 2005, 37).

There are other social problems that seem to be ignored in the analyses of Ho and Nurse (2005). The Caribbean "model" of tourism, according to Patullo (1996), tends to leave the local worker disadvantaged in terms of income and employment and encourages expatriate managers with the clichéd ideas about the Caribbean and its inhabitants to re-create small pockets of colonialism and all the attendant resentments and frissons (Patullo 1996: 65–70). The *Sunday Guardian,* in 2006, published a letter written by six prominent Tobagonians (including a former head of the civil service, Reginald Dumas) to the government which advised that the sale of lands to foreigners in Tobago was having deleterious effects on the locals of that island and should be stopped (*Sunday Guardian* August 27, 2006, 21).

Related to the effect of tourism on natives is the effort to create a "brand." The response to the presence of the Trinidad Carnival contingent at the 2006 Football World Cup illustrated the expectations cultivated of foreign tourists by the Carnival promotion. An article written by a European journalist featured an opinion from FIFA's president:

"Oil drum music is infectious . . . " says Sepp Blatter, the president of FIFA, organizer of the 2006 World Cup in Germany in June. Blatter envisions the rum poured and a conga line ensuing around the 10,000 steel-drum "pan men" expected to follow the Trinidad and Tobago Calypso Carnival Warriors [sic] team (*Trinidad Express* April 2, 2006, 22).

The concern with the relegation of locals to being colourful props is part of a well-established litany of tourist-industry ills, and for all the desire to create a special brand of tourism, the Germany effort might have merely confirmed existing stereotypes of rum-drinking, music-making natives.

THE NEW TOURISM IN TRINIDAD

The invitation to entertain assumptions of primitivism and fantasy fulfilment in Trinidad is not pitched to the sun-seeking tourist alone. Discussing her organization's activities, the director of a Trinidadian arts organization, Caribbean Contemporary Arts (CCA7) (which is a beneficiary of several international organizations, like the Prince Claus Fund, and the Carnegie Mellon and Ford Foundations), described the responses of foreign artists who visited via its programmes:

[The artists] came here with an absolutely clear idea of the kind of work they would do, and they all ended up doing something completely different once they'd had their first roti and win' in a rumshop. Noritoshi Hirakawa [a Japanese video artist] did obsessive research on a temple where they mix orisha [local voodoo] [sic] deities with Hindu ones.

Priests chanting people catchin' the spirit—it made a beautiful video (Darwent 2003).

A British artist, Chris Ofili, who participated in CCA7's programmes, saw another dimension of the island, as related in an article in the *Guardian* in 2002. After a visit to Trinidad, he was amazed and moved to find these values [of Black nationalism] alive and well.

> "I actually had the dry cleaner's image for quite a while, but it wasn't until I'd been there that it triggered the possibility of how to connect it with ideas of Afro love and Afro unity. Those kind of old-fashioned values still exist" (*Guardian* June 15, 2002).

The effect on these artists—seeing exotic and ethno-nationalist paradises— is similar to the effects the state-sponsored tourist programme encourages, and suggests that the discourse of "tourism nationalism" described previously is not merely cultivated in the populist/commercial sphere but in the academic and "high" cultural spheres. A necessary element of this programme is the academic endorsement of the intertwined discourses of nationalism, commerce, and culture, which amounts to a solemnization of Carnival culture.

The first phase of this solemnization was academia's recognition of Carnival and local culture as what Michel Foucault calls "subjugated knowledges," and its agents were the already-identified "brown bourgeoisie" of Trinidad. As Rohlehr notes, from independence, "The 'Creole' intelligentsia has defined both 'nation' and 'culture' in terms of 'Creole' imperatives to the exclusion of other ethnic groups" (Rohlehr 2004:142). Derek Walcott, present to observe the programme in action, described a "new, brown meritocracy," who, with the intellectuals, "apotheosised the folk form, insisting calypsos were poems" (Walcott 1998:30–31).

In discussing how conquest and its instruments ("institutions, practices and discourses") relate to the status of knowledge by the imposition of "totalizing theories," Foucault describes an "insurrection of subjugated knowledges" which may result in the "union of local and erudite knowledges." The combination of the totalizing, conquering knowledges like those of colonialism, and the local, "low-ranking" knowledges tend to create new genealogies. These genealogies, according to Foucault, "entertain the claims to attention of local, discontinuous, disqualified, illegitimate knowledges against the claims of unitary body of theory which would filter and order them in the name of some 'true' knowledge and some arbitrary idea of what constitutes a science and its objects" (Foucault 1980:83). The principal opposition here is of the "high," hierarchical colonial knowledge against the "low" local knowledge.

The emergence of Carnival as a "high-ranking" knowledge, though its "acceptance" started at independence, culminated only in the 1990s. The symbolic validation of the local knowledges was the award of honorary doctorates from the University of the West Indies to Peter Minshall, Pat Bishop, and Francisco Slinger (The Mighty Sparrow): a Carnival designer, steel-drum arranger and artist, and calypsonian, respectively. This signified the society's, state's, and academia's acceptance of the Carnival-nationalist worldview.

A crucial issue is the link between university academics, cultural authorities, and the brown bourgeoisie. The social basis of this relationship is to be found in Trinidad sociologist Acton Camejo's study, which described the independence-era networks of the business and social elites through clubs and associations, which demonstrated a high degree of intermixing among business, politics, and social-cultural organizations (Camejo 1971). These networks persist, albeit in modified form. For example, local academics are today habitués of social circles which were once the province of expatriates and the White elites described by Calder-Marshall in 1939. The extent of the commingling among academia, the brown bourgeoisie, and the business elites was illustrated in the Galvanize arts programme, which was organized as a "fringe" festival to Carifesta in September 2006. The organizing board of Galvanize consisted of the director of CCA7, journalists, university academics, artists, professionals, expatriates who occupy influential positions locally (like publishers and editors), and metropolitan academics.[8] Its relation to Carifesta was best articulated by a newspaper editorial which described "a group of young, cutting edge local artists" against "the elders charged with managing Carifesta IX" (*Trinidad Guardian* August 28, 2006, 28). The project received much media attention in print, broadcast, and quasi-academic media.

The Galvanize conflict was more than an ironic reenactment of the dialectical confrontation between "local" and "colonial" knowledges around independence, with the local, in this case, in the superordinate position. The contest between the "Creole" and colonial knowledges at independence created a binary opposition which suggested that those poles were the only cultural possibilities. And the reconstitution of the conflict between colonial knowledge and local knowledge signals the survival of the limited locus of possibilities at the expense of other knowledges.

Of these knowledges, certainly, the Indian cultural knowledge was and continues to be suppressed, but Indo-culture in Trinidad seems indifferent to recognition by or attention from state culture. It is a self-sustaining entity: its music and cultural events are widespread and widely patronized—locally and internationally (Ramnarine 2001; Balliger 2005a)—but with the exception of a few tokenistic inclusions, it is still largely ignored in national cultural collages.

The consequence of this closely managed cultural dialectic is the suppression of the *possibilities of cultural fusions and interactions* of European, Asian, and African cultures in Trinidad and Tobago. Essentially, the obsession with ethnicity, seeded and encoded into the cultural, academic, and political discourses, may be negating *unforeseen possibilities* in new modes of cultural expressions.

CONCLUSION

This chapter proposes that culture and cultural events have been used by the state apparatus mainly as national instruments in the destination "brand" of Trinidad and Tobago. Ostensibly this seems to be for the purposes of attracting tourists and building a favourable brand image. But Trinidad and Tobago's oil wealth frees it from a reliance on tourism earnings; and the limited investment in the development of the industry by the state, compared to Caribbean neighbours, suggests that this sector may not be a government priority. As such, the highly visible pursuit of tourism via the staging of events and cultural festivals require a more focused policy of promoting and representing the rich diversity and range of the ethnic groups in the country. Crucially, this will provide the double benefits of enhancing the appeal of the destination and may also engage the support of all local ethic groups in tourism development.

NOTES

1. The term *Creole* is a highly problematic one in local cultural matters. On one hand, it is promoted as an all-inclusive concept, a description of a dynamic process of acculturation and assimilation of cultures which encounter each other hierarchically. In practice, it is usually taken to mean a particular Afro-Caribbean dialectical process, where "foreign" things are "taken in hand" and oriented to the Afrocreole discourse. An excellent discussion of these issues is to be found in the collection *Questioning Creole: Creolisation Discourses in Caribbean Cultures* (eds. Verene Shepherd and Glen Richards), Kingston/London: Ian Randle Publishers/James Currey, 2002.
2. The BET documentary aired on BET on September 12, 2004.
3. The government of Trinidad and Tobago's official Web site is www.visittnt.com.
4. Data are yet unpublished but provided by Central Statistical Office: www.cso.gov.tt.
5. The PNM lost power in 1986 to a coalition (the National Alliance for Reconstruction [NAR]); the difference with 1995 was that the 1986 party was a multiethnic coalition which dissolved soon after it acquired office. In 1995, however, the PNM lost to what it most feared: an Indo-based party, the United National Congress.
6. *Divali*, Vol. 7, No. 2, Port of Spain: Indo-Caribbean Cultural Council, 2005, edited by Dr. Kumar Mahabir
7. I asked Nurse (publicly, during a panel discussion) at the Society for Caribbean Studies Conference in Warwick, 2002, whether he factored in things like the costs of AIDS and diversion of resources from other areas, such as national security, to combat Carnival violence, in his benefit-cost analyses, and he answered in the negative. He said he was not aware of any link between Carnival and violence.
8. Details of Galvanize can be found at projectgalvanize.blogspot.com

REFERENCES

Balliger, R. 2005a. Chutney soca music in Trinidad. In *Globalisation, Diaspora and Caribbean popular culture*, C. Ho and K. Nurse, eds. Kingston: Ian Randle Publishers.

———. 2005b. Hypernationalist discourse in the Rapso Movement in Trinidad and Tobago. In *Globalisation, Diaspora and Caribbean popular culture*, C. Ho and K. Nurse, eds. Kingston: Ian Randle Publishers.

Calder-Marshall, A. 1939. *Glory dead*. London: Michael Joseph Ltd.

Camejo, A. 1971. Racial Discrimination in employment in the private sector in Trinidad and Tobago: A study of the business elite and the social structure. In *Social and economic studies*. Mona: Institute of Social and Economic Research.

Darwent, C. 2003. Art review *Art Review Inc* LIV:37–8.

Foucault, M. 1980. Two lectures in power knowledge. In *Selected interviews and other writings, 1972–1977*, C. Gordon, ed. New York: Pantheon.

Higman, B. 1998. *Writing West Indian histories*. London: Macmillan.

Ho, C., and K. Nurse. 2005. *Globalisation, diaspora Caribbean and popular culture*. Kingston: Ian Randle.

International Monetary Fund [1], Public Information Notice No. 05/159, 2006, retrieved from www.imf.org on November 20, 2006

International Monetary Fund [2] Country Report on Barbados, 2006. Retrieved November 20, 2006, from www.imf.org.

Lewis, G. 2004. *The growth of the modern West Indies*. Kingston: Ian Randle Publishers.

Patullo, P. 1996. *Last resorts: The cost of tourism in the Caribbean*. Kingston: Ian Randle Publishers.

Ramcharitar, R. 2005. *Breaking the news: Media and culture in Trinidad; Port of Spain*. Port of Spain, Trinidad and Tobago: Lexicon.

Ramnarine, T. 2001. *Creating their own space: The development of an IndoTrinidadian musical tradition*. Mona, Jamaica: The Press, UWI.

Rohlehr, G. 1992. *My Strangled city and other essays*. Port of Spain, Trinidad and Tobago: Longman.

———. 2004. *A scuffling of islands: Essays on calypso*. Port of Spain, Trinidad and Tobago: Lexicon.

Shepherd, V., and G. Richards, eds. 2002. *Questioning Creole: Creolisation discourses in Caribbean cultures*. Kingston, Jamaica: Ian Randle Publishers.

Van Koningsbruggen, P. 1997. *Trinidad Carnival: A quest for national identity*. London: Macmillan.

Walcott, D. 1998. *What the twilight says: Essays*. London: Faber & Faber.

World Travel and Tourism Council (WTTC).Trinidad and Tobago: The Impact of Travel & Tourism on Jobs and the Economy (2005). Retrieved November 16, 2006, from www.wttc.org.

World Travel and Tourism Council (WTTC) League Tables: Travel and Tourism. Retrieved November 18, 2006, from www.wttc.org

Yelvington, K., ed. 1993. *Trinidad ethnicity*. London: Macmillan.

6 A Postcolonial Interrogation of Attitudes toward Homosexuality and Gay Tourism
The Case of Jamaica

Donna Chambers

Hunting batty bwoys[1] is as instinctive as the craving for fry fish an bammy, a national dish. The mere sight of them can trigger the bedlam of a witch hunt. When the toaster rapper (Hammer) Mouth discovers two gay men in a garage—"hook up an ah kiss like . . . mangy dog,"— he hollers: Run dem outa di yard. Murder them, advises another toaster, kill them one by one. Murder dem till dem fi change dem plan (Noel, 1993, cited in Crichlow, 2004:197)

INTRODUCTION

Gay tourism can arguably be seen as an increasingly lucrative form of niche or 'alternative' tourism because of the growing economic wealth of the gay traveller and the increasing legitimisation of homosexuality, particularly within the developed world. The apparent importance of the gay consumer in the context of travel and tourism has been suggested by several authors such as Hughes (1997), Clift and Forrest (1999), and Alexander (2005), who note that travel and tourism are an integral part of Western gay culture and identity. However, against this background it is surprising that to date there has been a dearth of studies on homosexuality and tourism. The earliest writings on tourism and homosexuality emerged a little over a decade ago with works including that by Clift and Wilkins (1995), Holcomb and Luongo (1996), Hughes (1997), Clift and Forrest (1999), and Pritchard, Morgan, and Sedgely (1998). Undeniably, for researchers in tourism who seem to be already unwilling to address issues of sexuality and corporeality, the subject of homosexuality, even today, seems to present an added complexity and sensitivity. The lack of studies in this area might also lie in the difficulties associated with definitions of homosexuality and consequently with identifying the gay traveller as a distinct market segment (Hughes, 2002).

Even so, where studies do exist on the gay tourism consumer, most of these focus on the gay consumer in terms of motivations for travel, demographic characteristics, and the destinations which are popular with the gay consumer. In the mainstream tourism literature, very little has been published on the attitudes of host societies towards homosexuality and travel (except for a limited discussion by Want, 2002). Thus, in this chapter I will seek to, in some way, contribute to the limited discussion on homosexuality and travel through a focus on the attitudes of host societies towards homosexuality, and by extension, towards gay travel, with particular focus on the Caribbean island of Jamaica. The chapter presents an exploratory argument which is grounded in a postcolonial theoretical context and which draws on evidence from an eclectic mix of secondary sources including books, journals, newspaper articles, the Internet, and magazines. Two issues are explored in this chapter: The first interrogates the argument that the negative attitudes of many Jamaicans towards homosexuality are a reflection of a wider postcolonial political struggle. The second is related to the extent to which it can be argued that the pressure exerted on the island to conform to the more 'liberal,' 'enlightened' attitudes of the developed, capitalist world towards homosexuality can be viewed as a form of postcolonial imperialism, which is partly being played out in and through the tourism industry.

Before commencing on this postcolonial exploration of the attitudes of Jamaicans toward homosexuality and, by extension, gay travel, it is imperative that I declare, from the outset, my 'situatedness' or location within this research project. I am a Jamaican heterosexual woman who has been educated predominantly in the Caribbean. I completed my doctoral studies in the United Kingdom and have remained in the country as a lecturer in tourism management. I have specifically identified myself as a Jamaican heterosexual woman because I believe that these 'subject positions' have influenced my choice and discussion of this topic. That is, my 'Jamaican-ness' means that I have intimate knowledge of the attitudes and behaviours of many Jamaican people toward homosexuality, which, admittedly, have coloured my own views on this topic. My identification as a heterosexual woman implies, on the other hand, a certain distancing from homosexual norms and behaviours (if one can identify such distinctions!), which perhaps limits my 'authority to speak' on homosexuality, particularly as it relates to gay males, who are the subjects of this chapter.

Yet as an individual from a country which experienced British colonialism for centuries, I was inspired to research this issue in 2004, which is a year when, arguably, the debate on Jamaica's attitude toward homosexuality received widespread negative coverage in the UK press.[2] Reading the many newspaper articles on this issue, it seemed to me that Jamaica was represented as an uneducated and uncivilised society which needed to be 'instructed' in enlightened behaviour by the former colonial power. Given that Jamaica had received political independence from Britain in 1962, it was axiomatic that the latter could no longer issue legislative directives on the discourse and practice of homosexuality and gay travel in Jamaica. Thus, a more covert strategy needed to be adopted to 'bring Jamaica in line,' and it seemed evident that one of the island's key export industries, tourism, was selected as one of the main tools that would be used in this civilising process. I must admit that I felt that Jamaica was being vilified for its attitudes towards homosexuality with little thought given to any underlying explanations for these attitudes. I believed, further, that the impression given in the UK press was also that there was no correlation between Jamaica's attitude toward homosexuality and its postcolonial condition. I believed that there existed a link between postcolonialism, homosexuality, and tourism which implicated both former coloniser and colonised and that this link should be explored. And that is the fundamental aim of this chapter.

A Word on Reflexivity

The discussion has commenced with a statement on my 'situatedness' or location within the research because this is intrinsic to the project. Indeed, the analyst cannot be separated from the phenomenon being analysed. This has often been described as an emic approach to research, which means that the research takes account of the insider's view of reality whether this is that of the subjects of the research or the researcher herself. This is often contrasted with a positivistic etic approach in which researchers are seen as standing outside of, and divorced from, the realities that they seek to investigate. In support of an emic approach to research, Said (1995:10), observes:

> No one has ever devised a method for detaching the scholar from the circumstances of life, from the fact of his [sic] involvement (conscious or unconscious) with a class, a set of beliefs, a social position, or from the mere activity of being a member of society . . . the general liberal consensus that "true" knowledge is fundamentally non-political (and conversely that overtly political knowledge is not "true" knowledge) obscures the highly if obscurely organized political circumstances obtaining when knowledge is produced.

Charmaz and Mitchell (1997:193) note that in positivistic, etic approaches: 'Scholarly writers have long been admonished to work silently on the sidelines, to keep their voices out of the reports they produce, to emulate Victorian children: be seen (in the credits) but not heard (in the text).'

It is within this context that the question of reflexivity has been addressed from the outset. Reflexivity means that I must not only make explicit the methodological position from which I am theorising, but also I need to declare and to reflect critically upon my own role within the research (Gill, 1995). It is axiomatic that researchers are active participants in the research process and as such reflexivity points to the need to understand the researcher's location of self (e.g., in terms of class, race, gender, ethnicity, citizenship, ideology, etc.; Hertz: 1997). Henwood and Pidgeon (1993) indicate that reflexivity implies that the researcher and the researched are interdependent in the social process of research. Borrowing from Rabinow (1986), cited in Michalowski (1997), it is necessary to understand that reflexivity is rooted in the awareness that the researcher, as a positioned subject, constructs interpretations of the 'facts' discovered through research. Reflexivity is necessary because it opens 'the way to a more radical consciousness of self in facing the political dimensions of fieldwork and constructing knowledge' (Callaway, 1992, cited in Hertz, 1997, viii). Research is therefore not a neutral, impartial activity, and while postcolonial theory supports a certain epistemological scepticism, this should not imply an eschewing of the question of values, particularly my own values (Gill, 1995).

So that my claim is that this study is reflexive, but not in an ethnographic sense where the researcher is engaged with 'live' subjects (individuals) but in the sense that the researcher is cognisant that she is an active participant in the way in which the research process has been conducted, the theoretical approach that has been adopted, and in the particular interpretations that are brought to bear on the materials that she is dealing with. Indeed, while I am not involved in an interactive way with individual subjects, I am nevertheless interacting with textual materials (newspaper articles, journals, books, magazines, Internet). So the specific interpretation of the attitudes of Jamaicans toward homosexuality, and, by extension, gay travel provided in this investigation is therefore necessarily influenced by my own values and *situatedness* or location within the research. In other words, I am essentially a 'situated actor.' With this in mind, it is important to note that this research is reflexive in a dual sense.

First, it is reflexive in terms of the theoretical approach that has been adopted and that has influenced the research question to be investigated, the research process undertaken, the data collected for what is essentially an exploratory study, and the analysis and conclusions arrived at. This reflexivity, which is defined as 'methodological situatedness' (Chambers, 2003), lies in postcolonial theory, and this has explicitly pervaded the discussions in this chapter and has influenced the interrogation of Jamaica's attitudes toward homosexuality. Second, it is reflexive in terms of the particular interests, values, and preconceptions that are brought to the research. This reflexivity, which is defined as 'subjective situatedness' (Chambers, 2003), has implicitly pervaded every aspect of the research process but especially in terms of the particular meanings and understandings brought to bear on the analysis. In this context, the 'voice' of the researcher is thus inherently 'heard' in the research.

Having explained my dual 'situatedness' (i.e., both methodologically and subjectively) within the research, it is now necessary to explain how the discussion will proceed. I will commence with a very general overview of the key theoretical underpinning of this exploratory journey, that is, postcolonialism. I will proceed to discuss the concept of homosexuality and will then demonstrate the link between homosexuality and tourism. This will be followed by a discussion of the attitudes of Jamaicans towards homosexuality and, by extension, gay travel. The chapter will conclude with some reflections and a suggestion of a possible way forward.

EXPLORING THE CONCEPT OF POSTCOLONIALISM

The publication of Edward Said's seminal text *Orientalism* in 1978 can, arguably, be said to mark the formal emergence of colonial discourse as a defined area of study within academia (Williams and Chrisman 1994). Colonial discourse analysis sought to critique the various techniques, strategies, and processes by which knowledge about the 'Other' was constructed and produced. Colonial discourse analysis sought to interrogate and indeed to deconstruct the ways in which the colonial subject was seen and understood by the former colonisers and the power/knowledge structures and relationships inherent in these understandings. However, undoubtedly, many of these critiques of the colonial condition emerged, in a chronological sense after or *post* the colonial era. More importantly, they investigated the continued hegemony of colonial political, economic, and cultural discourse and practice on postcolonial societies and, for these reasons, might best be deemed as *postcolonial* theorising or postcolonialism. D'Hauteserre (2004:235) notes that postcolonialism can be 'defined as reflexive Western thought, interrogating and rethinking the very terms by which it has constructed knowledge through the duality of coloniser and colonised.' A key observation of postcolonial theory is that

> Colonial and imperial rule was legitimized by anthropological theories which increasingly portrayed the peoples of the colonized world as inferior, childlike or feminine, incapable of looking after themselves (despite having done so perfectly well for millennia) and requiring paternal rule of the west for their own best interests . . . The basis of such anthropological theories was the concept of race. In simple terms, the west-non-west relation was thought of in terms of whites versus the non-white races. White culture was regarded (and remains) the basis for ideas of legitimate government, law, economics, science, language, music, art, literature—in a word, civilization (R. Young 2003:2–3).

In other words, postcolonial theory argues that the relationship between the coloniser and the colonised was contextualised and legitimised primarily in terms of gender and race. In terms of gender, the colonies were essentially feminised, and this had connotations of sexual, political, and economic subservience. The emergence of tourism as a postmodern phenomenon has merely continued this relationship as argued by Graburn:

> At a psychological level these nations (Third World Countries) are forced into the female role of servitude, of being penetrated for money, often against their will, whereas the outgoing, pleasure seeking penetrating tourists of powerful nations are cast in the male role (Graburn, 1983, cited in Oppermann, 1999:253).

According to Sinha, 1995 (cited by d'Hauteserre 2004:239), this 'masculine imperialism' led to the feminisation (and associated 'inferiorisation') of the colonised male. Yet it seems rather paradoxical that at the same time that colonialism 'emasculated' the black male through its feminisation of the colonies, it perpetuated a mythology of a black male sexual threat to white femininity. This myth of the black male as sexual predator points to the racialised discourse of colonialism, which sees the nonwhite races as primitive and culturally and socially inferior.

Further, colonialism perpetuated the hierarchical dualism between mind and body in which the colonising power was seen to embody characteristics associated with the mind such as reason, subject, consciousness, and, importantly, as indicated previously, masculinism. On the other hand, the colonies were seen to embody characteristics associated with the body such as passion, object, and, indeed, feminism. Within the context of these racialised and gendered binary oppositions between coloniser and colonised, the latter had no authority to speak and must therefore depend on the colonisers to instruct on the 'correct' ways of seeing and being. Fundamentally, postcolonialism rejected these hierarchical dualisms between coloniser and colonised and, further, sought to elaborate a 'politics of the subaltern' (R. Young 2003:6). Postcolonialism, in this sense, was both a theoretical construct and a political call to action.

This necessarily brief and very general discussion of postcolonialism would be inadequate without the introduction of some critiques of postcolonialism itself. McClintock (1994) has argued that postcolonialism, while it seeks to eradicate the binary oppositions created by colonialism (primitive/modern, metropolis/periphery, self/other, etc.), itself 'reorients the globe once more around a single, binary opposition: colonial/post-colonial' (1994:292). She continues by arguing that postcolonialism seeks to recentre 'global history around the single rubric of European time. Colonialism [thus] returns at the moment of its disappearance' (McClintock, 1994:293). Postcolonialism, then, represents a singularity of vision, thus occluding the multiplicity of postcolonial conditions[3] with their diverse power imbalances. For Castle (2001), cited in d'Hauteserre (2004:236), 'postcolonialism remains complicit with colonialism due, in great part, to its origin in the Western intellectual tradition.'

The question that needs to be asked here is whether the 'post' in postcolonialism represents a radical rejection of colonialist modes of thought and structures only to replace these with another hegemonic discourse. While there is much merit in this critique of postcolonialism, it can scarcely be denied that postcolonial theorising has provided a necessary framework within which colonial and, indeed, imperial systems of domination can be critiqued. Further, it is evident that in the long trajectory of African, Asian, and European history, colonialism represents a pivotal moment which dramatically altered the relationship between and amongst these groupings. In this chapter I have focused my discussion on Jamaica, in a case study approach, which is a useful technique that can be adopted to highlight a specific postcolonial experience, which, while it might demonstrate similarities with the experiences of other Caribbean ex-colonies, should not be used to generalise on a wider scale.

It is important to note at this point that a theory of postcolonialism, by rejecting hierarchical colonial dualisms, also necessarily recognised that the prolonged history of colonialism (and here this is especially relevant within the Jamaican context, where this condition was experienced for over 400 years) inevitably led to some intermingling of the cultures of both coloniser and colonised (granted that the former was the more dominant), creating a kind of ambivalence on both sides. Indeed, one might argue that what exists in postcolonial societies is a 'complex mix of attraction and repulsion that characterises the relationship between coloniser and colonised. The relationship is ambivalent because the colonized subject is never simply and completely opposed to the colonizer' (Ashcroft, Griffiths, and Tiffin 1998:12).

Homi Bhabha, in his text the *Location of Culture,* had argued previously that cultures are not as distinct as is often thought but that many people, especially in postcolonial societies, exist in 'third spaces' which are located 'in between' areas of seeming difference (Bhabha 1994). Bhabha was no doubt influenced in this notion of ambivalence by Lacanian[4] psychoanalysis and Saussurean[5] structuralist theory. These latter theorists argued in different ways that meanings and identities are defined by what they are not. What is external to a particular meaning or a particular identity, what lies outside, is thus integral to the constitution of that meaning or that identity. In other words, identity and meaning are established through difference. What is external to a meaning or an identity is known as the constitutive outside, an 'absent presence,' or in Lacanian terms the identity's 'lack.' Stuart Hall, in a discussion in a publication by Campbell and Tawadros (2001), reflects on this idea:

> I do not know of any identity which, in establishing what it is, does not at the very same moment, implicitly declare what it is not, what has to be left out, excluded . . . We understand sameness only through difference, presence through what it 'lacks.' To say or establish anything—any position, any presence, any meaning—one has to attend to what is outside the field of meaning and what cannot be expressed—it's constitutive outside. The truth of the Lacanian insight is that the subject is constructed across a 'lack,' the self by its 'others.' . . . Difference then is not something that is opposed to identity; instead, it is absolutely essential to it (Campbell and Tawadros 2001:40–41).

What this means is that colonial discourse in establishing the colonised as 'other' often fails to recognise that the existence of the colonised is integral to the coloniser's sense of identity. However, the condition of colonialism, as it existed in Caribbean societies like Jamaica, problematised the distinction between the inside and the outside, thus sometimes resulting in a blurring of the lines between coloniser and colonised. The onset of mass immigration from the Caribbean to the United Kingdom in the mid twenteith century further disrupted the divide between coloniser and colonised. This recognition of the existence of a kind of confluence between coloniser and colonised, I suggest, is often ignored in the polemic on homosexuality and gay tourism as they relate to Jamaica.

TOURISM AND HOMOSEXUALITY

While undoubtedly men have been having sex with other men since time immemorial, the *concept* of 'homosexual' has been deemed by Foucault and other historians as an invented social identity which emerged in the 1860s and the subsequent decades (Pinar 2003). Indeed, according to Foucault, 'Homosexuality appeared as one of the forms of sexuality when it was transposed from the practice of sodomy onto a kind of interior androgyny, a hermaphrodism of the soul. The sodomite had been a temporary aberration; the homosexual was now a species' (Foucault 1984:322–23).

Howard Hughes, who is, arguably, one of the earliest and most prolific writers within the mainstream tourism literature on homosexuality and tourism, supports this notion when he locates the birth of homosexuality as a 'source of identity rather than a sexual act' in postindustrial society when 'male and female roles became more sharply distinguished' (1997:4). Hughes continues by stating that 'a homosexual identity is a choice though it may have some other, perhaps, biological basis' (1997:4). While it is difficult to agree on the 'causes' of homosexuality, it is hard to disagree that homosexuality is a social identity and, in Western society, being gay (and here the focus is predominantly on the male homosexual and to a lesser extent on female homosexuals or lesbians) constitutes a postmodern lifestyle with the gay consumer becoming a sought-after market segment, particularly within travel and tourism.

A report by Mintel International, published in 2000, on the gay travel consumer in the United Kingdom noted that they have a much higher rate of holiday taking than the rest of the travelling population, they have a higher than average personal disposable income (PDI) due to the lack of dependents, and they are in a higher socioeconomic group. In 2005 the Travel Industry Association of America reported that the American gay and lesbian community represented a US$65 billion travel market, which amounted to about 5 percent of the annual US$1.3 trillion travel industry (*Travel Wire News* November 2, 2005). In the United Kingdom, the gay traveller is seen as a lucrative market segment such that in 2006 VisitBritain, which is the official tourist agency for the United Kingdom, launched a marketing campaign that was aimed specifically at the gay travel consumer. In its Internet advertising message, VisitBritain declared: 'Welcome to the United Queendom of Great

Britain . . . with our proud gay history, cutting edge culture and fashion, flamboyant cities and pulsating nightlife, isn't it time you came out . . . to Britain?' (David 2006).

As indicated previously, it has been argued that there is a strong relationship between homosexuality and tourism:

> The acceptance of a homosexual identity is often dependent upon the act of being a 'tourist' at least in the limited sense of travel if not stay . . . The gay man is, in large part, able to be himself only in the gay space which may be primarily a leisure environment. . . . In addition, many gays will choose to travel in search of an anonymous or safe environment in which to be gay. (Hughes 1997:5)

This notion of the quest for gay space as a motivation for travel is confirmed by Clift and Forrest (1999) in their study of gay men in England, and Pritchard, Morgan, and Sedgely (1998) point to the safety that gay men feel when they occupy gay spaces largely within the context of leisure. According to Mintel International (2000), what gay travellers seek is gay friendliness and understanding in their holiday destinations rather than a specifically gay-themed holiday. This idea of the need for gay spaces and gay friendliness while on holiday seems to axiomatically exclude destinations like Jamaica, where public gay spaces are conspicuous in their absence. It is my contention that the attitudes of a host country toward homosexuality would necessarily point to the extent to which that host country would seek to create and promote public gay spaces within the context of travel and tourism.

Indeed, in the Caribbean and in Jamaica, the subject of homosexuality is still largely taboo. Lewis (2003) has indicated that in the Caribbean sexuality is only openly discussed and explored within the context of popular discourse. He suggests that there is a 'strong correlation between sexuality and popular culture' (2003:7). Outside of this context, debate on sexuality, particularly that of homosexuality, is scarcely heard. In Jamaica, where discussions of sexuality and corporeality take place, these discussions are limited to heterosexuality with the concomitant marginalisation of homosexuality. However, in contrast to what obtains in the wider society, in popular culture, particularly in dancehall music, the subject of homosexuality (especially as it relates to gay men) receives significant attention. In the next section the prevailing attitudes in Jamaica toward homosexuality and, necessarily by extension, gay travel will be discussed.

JAMAICA, HOMOSEXUALITY, AND TOURISM: AN UNWELCOME TRIUMVIRATE?

The contention of this chapter is that in Jamaica attitudes towards homosexuality are grounded not only in the cultural context of the island but also in its legal and religious systems. The Offences Against the Person Act Section 76 provides for a prison term not exceeding 10 years for 'whosoever shall be convicted of the abominable crime of buggery, committed either with mankind or with any animal,' while Section 79 describes as 'outrages on decency' the commission 'by any male person of . . . any act of gross indecency with another male person.' It is interesting to note that the law is specifically against buggery (commonly understood as anal intercourse) predominantly between males and not, theoretically, against homosexuality as a social identity. The focus is therefore on criminalising the sex act largely believed to be carried out between gay men. One might assume, therefore, that Jamaicans cannot legally be prosecuted for being homosexual per se. Importantly, this act is a legacy of the colonial era (and here I am referring to the 1861 Offences Against the Person Act in the United Kingdom, which was also promulgated in its colonial possessions). Indeed, it should be recalled that it was only in 1967 that the Sexual Offences Act was passed in the United Kingdom, which only *partially* decriminalised sex between two men (as long as it was consensual, done in private, and both parties were over 21 years old; BBC News 2002). Further, according to Alexander (2005:48), these sodomy laws had racist and heterosexist underpinnings and were introduced by the British in their colonies at a time when colonial rule was being consolidated, with these laws representing one of a number of potential offences against 'the white, male British person.'

In religious terms, Jamaica is predominantly a Christian society, with one of the highest densities of churches per square kilometre than anywhere else in the world, and in this context homosexuality is seen as being contrary to biblical tenets. Specifically, Jamaicans still point to the following biblical passages as support for the 'sinfulness' of homosexuality. The first is the instruction given to Adam and Eve in Genesis 1:28, which mandates them to 'Be fruitful and increase in number,' which command can be carried out only through male/female reproduction. The second is the story of Sodom and Gomorrah in Genesis 19, which was allegedly a city largely inhabited by homosexuals and thus destroyed by God for this iniquity. The third passage is Leviticus 20:13, which states that 'If a man lies with a man as one lies with a woman, both of them have done what is detestable. They must be put to death: their blood will be on their own heads.' Again, Christianity is another legacy of colonialism and was further a key pillar of colonial society used by the colonial power to ensure the continued passivity and servitude of its subjects.

However, I suggest that it is in culture and specifically in popular culture that attitudes toward homosexuality are most overtly discerned. In this regard, the argument is that cultural attitudes range from a qualified tolerance to outright hostility. An example of the former is in this statement by a prominent Jamaican newspaper columnist: 'Like the vast majority of Jamaicans, I in no way condone violence against homosexuals or even discrimination against them . . . [but] Am I ready to cheer as gay parades with

some of the more unusual looking human beings go prancing by? Not Yet!'
(Abrahams-Clivio 2004).

Another example of this qualified tolerance is contained in a commentary
by Mark Wignall, a frequent writer in a popular Jamaican newspaper, who,
reflecting on the murder in 2004 of Brian Williamson, head of JFLAG[6] and
the most vocal advocate of homosexuality in Jamaica, declared:

> I do not want homosexuals trying to convince my grandson that 'it' is
> an alternative lifestyle. I don't want them writing books where a prince
> falls in love with another prince and they kiss and then live happily ever
> after. I have never burnt a book in my life, but there is always a first for
> everything . . . most of us are living with homos in our midst, and if they
> keep their hands and desires to themselves then we are fine. We don't
> like them, but so what? They were given a bad hand at birth in this—an
> imperfect world. Hell, maybe they don't like us either. So what? There
> is space here for all of us. (Wignall 2004)

On the other hand, an example of an attitude of outright hostility to
homosexuals is contained in this extract from a letter to the editor of the
main newspaper in Jamaica: 'Regardless who is the spokesperson for homosexuality it cannot be right, it is a detestable act with shame and dire consequences to those who practice it' (Stennett, 2002).

John Maxwell, a noted Jamaican political commentator, observed in
2004 that

> Several years ago, the Jamaican press carried a story to the effect that
> homosexuals were planning a march from Half-Way-Tree to Jamaica
> House. The story, on the face of it, was outlandish, but that did not stop
> the agents provocateurs of the press from carrying it. And as they had
> hoped, dozens of machete wielding assassins gathered in Half-Way-Tree
> square on the appointed day to deal out their brand of justice to any
> homosexual who dared show his face. (Maxwell 2004)

However, some of the most vitriolic condemnations of homosexuality are
evident in Jamaican dancehall music, which is a key expression of Jamaican
popular culture. Dancehall music blossomed in Jamaica during the preinde-
pendence 1950s in Kingston's inner cities, characterised by colonial legacies of
poverty, overcrowding, and violence amongst a largely black male 'underclass.'
Within the context of this kind of deprivation, dancehall music, which was
largely the province of young black males, was used as a medium to express
an exaggerated masculinity in the face of the emasculation of the black male.
According to Stanley-Niaah (2004:103), dancehall music 'tells the story of a
people's survival and the need for a celebration of that survival against forces
of imperialism and systems of exclusion. Dancehall's story is ultimately the
choreographing of an identity that critiques aspects of Western domination.'

For dancehall artistes, many of whom emerged from Jamaica's inner cities, poverty was antithetical to masculinity. Through success as dancehall artistes, many of these young men were able to escape poverty and its attendant emasculation. I suggest that colonialism, and the poverty it spawned, so disrupted the boundaries between male and female that the homophobia displayed within the popular culture can be perceived as a means by which a male identity was reestablished.[7] Indeed, as argued by Young:

> Homophobia is deeply wrapped up in issues of gender identity . . . the genders are considered to be mutually exclusive opposites that complement and complete one another. Order thus depends on the unambiguous settling of the genders: men must be men and women must be women. Homosexuality produces a special anxiety then as it seems to unsettle this gender order. Because gender identity is a core of everyone's identity, homophobia seems to go to the core of identity. (I.M. Young 2003:381)

In fact, rather paradoxically, in Jamaica homosexuality is often perceived as a sexual deviance which has been imported and imposed from the West, thus representing a way of life that is nonindigenous to the island. Jamaica is thus caught 'in a contradictory homage to the West as the harbinger of civilization and progress on the one hand and as the harbinger of sexual destruction on the other' (Alexander 2005:49). By expressing such vitriolic anti-gay sentiments, not only are the dancehall artistes proclaiming their own manhood, thereby rejecting the feminisation of the colonial male occasioned by colonialism, but are also rejecting the dominance of Western cultural and social systems which sanction homosexuality. This resistance to Western domination is also expressed by the former prime minister of Jamaica, P. J Patterson, who, at a Party conference in 2000:

> Issued a warning to human rights organisations and overseas governments that despite their 'pressuring and condemning us, laws allowing capital punishment and those making homosexuality illegal, will be upheld in Jamaica' . . . he also told party delegates and observers that 'under my watch and under the Peoples National Party, we have no intention whatsoever of changing those laws that make homosexuality illegal.' (Virtue 2000)

Five years later, in 2005, in an interview with the UK *Guardian* newspaper in which he was asked to repeal the laws that made buggery illegal, then Prime Minister Patterson declared: 'Let me make it very clear, the laws of Jamaica must be determined by the Parliament of Jamaica, and that right we will maintain. We will never, never, compromise.' (Aitkenhead 2005)

This resistance to attempts by the former colonial power to infringe on Jamaica's political sovereignty was also previously reflected in the following statement by a then Jamaican Member of Parliament Ronald G. Thwaites:

> How wrong it is to have a British Court impose their moral standards on us . . . Even though I would like to see the people's opinion on the laws to do with hanging and homosexuality change one day, I recognise the need to have our legal institutions interpret and enforce the laws that our Parliament has approved. (Thwaites 2000)

However, I suggest that as the forces of globalisation become more powerful and omnipresent, Jamaica is increasingly being criticised for its failure to conform to recent advances within the developed world which are moving towards the legitimisation of homosexuality. Jamaica's economic export industries, specifically the vital tourism industry, but also the music industry, is facing sanctions from developed countries such as the United States and the United Kingdom for the country's attitudes towards homosexuality. In the case of the music industry, popular dancehall artistes are seeing their shows cancelled, the withdrawal of music awards, and even the refusal of physical access to the United States and the United Kingdom:

> In 2004, Elephant Man and Vybz Cartel were both nominated for best reggae act at the MOBO awards in the UK. However, their nominations were withdrawn in a furore about the homophobic lyrics in their songs which had titles such as Bedroom Slaughteration (by Vybz Cartel) and We Nuh Like Gay (Elephant Man) which were claimed to be advocating the murder of gay men. (BBC News, UK Edition, 8 September 2004)

And also in 2004 it was reported that

> Since July 30, 30 US concerts by Beenie Man have been cancelled, as have 7 US concerts by Capleton, a Scandinavian tour and four other European dates by Buju Banton, two UK tours and a San Francisco show by Sizzla, two European dates by Bounty Killer and a Reggae in the Park festival scheduled to be held at the Wembley Arena. (Petridis, 2004)

While these sanctions were directed against individual dancehall artistes, they will have implications for the wider music industry in Jamaica.

More importantly, the island's tourism industry, which has been the main foreign exchange earner since the 1980s and which is heavily dependent on tourist flows from developed countries, including the United States and the United Kingdom, is, I propose, also being used as one of the key tools to bring the country in line with current developed-world thinking on homosexuality. JFLAG reported that in the late 1990s, when both the government of Jamaica and of Cayman refused to allow gay cruise ships to dock, the executive director of the powerful International Gay, Lesbian and Transgender Association (IGLTA), based in the United States, stated that 'If Jamaica is unwilling or unprepared to welcome gay and lesbian tourists to its shores, then IGLTA is prepared to warn all of our member companies and associations that our tourist dollars are no longer welcome in that country' (cited in Alexander 2005:86).

Mintel International, a prestigious company that produces reports including those on UK industries such as travel and tourism, in a report on gay tourism published in 2000, indicated that

> there are a number of resorts such as Sandals, and countries such as Jamaica, which have a known record of homophobia which the industry needs to be aware of—thus avoiding possible situations of embarrassment or even danger for gay clients. Jamaica, for example, is known to be a particularly homophobic destination and some aggression towards gay holidaymakers has occurred in the past. The Sandals resorts have a policy of allowing 'heterosexual couples only' at their couples resorts and have been given the Rock Bottom award seven years running by Out and About—the gay travel information magazine. Many gay organisations have been lobbying the travel industry to stop selling Sandals but Microsoft's Expedia is the only company to date that has made a point of withdrawing Sandals from its supplier list. (Mintel 2000)

Sandals Resorts is a major Caribbean all-inclusive hotel chain which is based in Jamaica, where the largest percentage of its resort properties are located. Sandals was subsequently banned from placing television advertisements in Britain for 'failing to make clear that its resorts are not open to gay and lesbian couples' (Reuters 2001, cited in Want 2002:197). Three years later, Sandals was also banned from placing its advertisements on the London Underground due to its continued unwelcoming attitudes towards homosexuals. Threats were also made that its advertisements would be banned from London taxis. In the face of this economic pressure, Sandals was allegedly forced to officially withdraw its anti-gay policy:

> Sandals, the Caribbean resort company announced it was lifting its ban on same sex couples from 13 resorts . . . Sandals was under commercial pressure from London's mayor, Ken Livingston who had banned its advertisements from the tube because of its homophobic attitudes to clients. Mr. Livingston was seeking to extend the ban to London's taxis. Sandals has operated a ban since 2001 and has been criticised for it in parliament. (Hencke 2004)

Thus it is that the tourism industry, a key driver of the Jamaican economy, is being used as a tool to bring the island in line with the more 'liberal,' 'enlightened' attitudes of the developed, capitalist world towards homosexuality. I argue that it is in this sense that the [former] colonised are constructed as primitive and barbaric with the aim of the imperial project being to ensure cultural and moral improvement.

DISCUSSION

In light of the foregoing, it is evident that there is a strong link between the largely negative attitudes toward homosexuality in Jamaica and its colonial condition. Indeed, during the colonial period a number of laws were promulgated in Britain and its colonies which deemed homosexual intercourse as immoral and illegal and thus punishable by law. These laws have remained on the books in former colonial territories like Jamaica long after they were discontinued by the British. While the British were able to force their Caribbean dependencies (which are also dependent on tourism as an engine of economic growth), of the Cayman Islands, Montserrat, the Turks and Caicos Islands, and the British Virgin Islands to repeal their anti-gay legislation (Want 2002), they have been unable to, overtly, do the same in Jamaica, which obtained its political independence from Britain in 1962. Further, Christianity, which is believed to regard homosexuality as an aberration, punishable by God, was also initially imposed on the colonial population in an attempt to ensure their 'civilization' and their servitude.

Yet despite this evidence of a colonial connection, Britain has now distanced itself from the attitudes of Jamaicans toward homosexuality. Instead, Britain has in effect reinforced hierarchical colonial dualisms such as primitive/modern. Interestingly, it has, arguably, only been in the last two decades that there has been an increasing acceptance of homosexuality as the 'norm' within significant parts of the Western world, including the United Kingdom. In this regard, Mintel International (2000) has suggested that

> Tolerance of homosexuality in the UK has improved greatly over the last 30 years. The last 10 years in particular have seen a significant growth in an open gay culture and with it a gay economy. The so-called 'pink pound' became a marketing catchphrase of the 1990s when marketers realised the potential profit reward of this market segment due to the perceived higher PDI patterns of the gay population and different lifestyle patterns. *The UK has a long tradition of homophobia.* (author's emphasis)

Indeed, this relatively recent acceptance of homosexuality in the United Kingdom is still not upheld by significant proportions of the British population. In 2006 Hickman reported that 'Gay men and women still encounter widespread discrimination and violence in the streets and workplaces of Britain, according to the biggest ever survey of gay lifestyles in this country' (Hickman 2006).

Still, the more 'enlightened' attitude of the United Kingdom towards homosexuality and gay travel is juxtaposed with the continued demonisation of homosexuality in large tracts of the developing world like Jamaica, with the latter seen as uncivilised and unenlightened. Undoubtedly the economic importance of tourism and the gay consumer within this context has served to heighten this issue. Importantly, there is, in this attempt to portray postcolonial societies like Jamaica as inferior, a lack of recognition of the continuities between coloniser and colonised. In the context of gay tourism, Murray (2007), citing Giorgi (2002), has argued that the mainstream gay tourism industry, in their evaluations of gay destinations around the world, construct destinations in the Caribbean as being 'primitive' when juxtaposed with the more 'progressive' and 'civilised' developed world. He goes further to argue that gay tourism discourse is one of 'authority and witnessing that validates political progress, historical advances and dimensions of the visible in foreign lands' (Giorgi 2002, cited in Murray 2007:52). For Murray the gay tourist 'In a sense, becomes a flag bearer of a progressive, socially visible gay identity connected to enlightened democratic nation states who is then used as a standard in evaluating the progress of "non-western" societies' (Murray 2007:52).

To further illustrate this point, in 2006, PlanetOut Inc, a leading media and entertainment company based in the United States, which specifically targets the gay and lesbian community, conferred the award for best travel destination of the year to Spain at its 13th Annual Travel Awards. Commenting on the reason for this award to Spain, the travel editor of PlanetOut indicated that

> There might as well be a map of Spain next to the definition of 'progressive society' in the dictionary. With a headstrong leap into modernity, Spain has doffed its Franco past and embraced an inclusive social policy such as legalized same-sex marriage, hedonistic nightlife, cutting-edge cuisine and fast-moving energy amid balconies and Iberian charm. (*Travel Wire News* June 27, 2006)

The implication here is that travel destinations which are not 'gay friendly' (such as Jamaica) are nonprogressive and premodern as juxtaposed to their European counterparts.

Clearly, it is still the case that postcolonial countries which seek to resist the dominant ideologies of the former colonisers do so only at huge political and economic cost (witness, for example, the invasion of the tiny Caribbean island of Grenada by the United States in the 1980s). The imposition of economic 'sanctions' (through the tourism industry) on countries like Jamaica because of their refusal to accept the legitimacy of homosexuality can be seen as one of the more recent manifestations of this imperialistic relationship. Yet I argue that the imperial countries fail to see the continuities between colonial and colonised discourses and practices. They fail to see that the 'outside' is also 'within' as expressed in this comment by Aitkenhead:

> But the vilification of Jamaican homophobia implies more than a failure to accept postcolonial politics. It's a failure to recognise 400 years of Jamaican history, starting with the sodomy of male slaves by their

white owners as a means of humiliation. Slavery laid the foundations of homophobia, and its legacy is still unmistakeable in the precarious, over exaggerated masculinity of many men in Jamaica. . . . Every ingredient of Jamaica's homophobia implicates Britain, whose role has maintained the conditions conducive to homophobia, from slavery through to the debt that makes education unaffordable. For us to vilify Jamaicans for an attitude of which we were the architects is shameful. To do so in the name of liberal values is meaningless. (Aitkenhead 2005)

So that, paradoxically, while Jamaica is being criticised for its discourse (most obviously expressed through the legal system and popular cultural expressions) and practice (through the operations of its tourism industry) on homosexuality, the extent to which the former colonial power is itself implicated in this scenario is scarcely discussed.

CONCLUSION

In this chapter I have sought to establish a link between Jamaican attitudes towards homosexuality, which I suggest would preclude the official development of gay tourism on the island, and its postcolonial condition. My suggestion has been that the negative attitudes of many Jamaicans towards homosexuality (which ranges on a continuum from qualified tolerance to overt hostility) can be seen as a reflection of colonial legal and religious legacies. Popular cultural attitudes towards homosexuality in Jamaica, notably as expressed in dancehall music, are largely reactionary insofar as they reject the feminisation and widespread poverty of the island during colonialism, both of which resulted in black male emasculation. Hostility towards gays is thus one attempt to reestablish manhood. The continued criminalisation of the act of buggery through the legal framework is also fiercely defended as a right of political independence from colonial rule.

Indeed, since 1962 Jamaica has been a politically independent country with the right to its own self-determination. As an independent country, Jamaica has decided to uphold the laws against homosexual intercourse, to continue to subscribe to Christian biblical tenets, which allegedly perceive homosexuality as deviant behaviour, and to continue to speak out against homosexuality in its popular cultural expressions. However, the forces of globalisation and the increasing importance of travel and tourism within this context mean that Jamaica is increasingly being criticised by the West for its antigay stance. Attempts are being made to coerce the island to conform to a more enlightened Western way of thinking towards homosexuality through the placing of economic 'sanctions' on its music and, more importantly, its tourism industry. Colonial hierarchical dualisms of civilised/modern, self/ other are again coming to the fore in the relationship between ex-coloniser and former colonised. Yet I suggest that by presenting Jamaica as 'Other' in its attitudes towards homosexuality, Western countries like Britain are failing to recognise the blurring of the boundaries between self and 'Other.' Indeed, the acceptance of homosexuality is a relatively recent phenomenon in the UK and even here there are significant pockets of resistance.

However it might be argued that by focusing on the colonialism/post-colonialism dualism in explaining Jamaica's attitude toward homosexuality, I am also subscribing to a new hegemonic discourse, a single binary opposition of colonial/postcolonial which was one of the main criticisms of postcolonial theory put forward by McClintock (1994). It might also be argued that by locating Jamaica's antigay attitudes in the context of its postcolonial condition, I am thereby robbing the Jamaican government and people of agency and am, moreover, occluding the voices of gay and lesbian Jamaicans, who are already marginalised in the island. Admittedly, including the voices of gay and lesbian Jamaicans would add richness and an alternative perspective to this discussion and is in fact the aim of a future research project. With regard to the question of agency, I am not suggesting that Jamaicans should not take some responsibility for their own attitudes towards homosexuality. Rather, my fundamental point here is that it is necessary to see the continuities and points of intersection between coloniser and colonised in an understanding of Jamaican attitudes to homosexuality and, indeed, gay travel. It is important to see how the attempt by Western countries such as Britain to coerce Jamaica, through imposing 'sanctions' on key economic industries like tourism, is reminiscent of the power relationships that existed during colonialism and how this might thus be perceived by Jamaicans as a new form of imperialist domination. And it is important to see the paradox inherent in the way in which colonial dualisms of civilised/modern can become blurred.

My arguments are not moral judgements about the rights and wrongs of homosexuality. My arguments are also not about whether the views of Jamaicans on homosexuality can be considered as consistent with what is now deemed as 'progress' within an increasingly globalised world. Indeed, the continued recalcitrance of Jamaica with regard to the legitimisation of homosexuality and gay travel might well be inconsistent with a country that has so readily embraced a global industry such as tourism, now the island's main foreign exchange earner. Rather, what I hope I have presented is one explanation for Jamaica's largely antigay attitudes that is informed by postcolonial theory. Undoubtedly, further research will need to be undertaken to inform this discussion, and, as I indicated in my introduction, this chapter represents a journey of exploration.

Ultimately, Jamaicans themselves will need to change these attitudes without having such a change imposed in what appears to be a kind of imperialist 'enlightenment' project. Importantly, strategies to encourage change need to take into account those wider contextual issues mentioned in this chapter. As a Jamaican newspaper columnist has suggested, 'Certainly gays in other countries . . . should not expect us to bypass the same process they went through' (Abrahams-Clivio, 2004). Indeed, in order for gay tourism to be considered as an officially sanctioned option for Jamaica's tourism industry, the island must deal with its own attitudes towards homosexuality, which, importantly, also serve to marginalise its own gay population. An acceptance by Jamaica of homosexuality would then lead to the creation of gay spaces within which gay travel can take place. However, any development by Jamaica of gay tourism must itself seek to avoid the power relationships that existed during colonialism by positioning Third World gay men as the 'objects of sexual consumption rather than as agents in a sexual exchange' (Alexander 2005:79). Indeed, white gay capital should not be allowed to become 'an active participant in the same processes of nativisation and recolonisation that heterosexual tourism helped to inaugurate' (Alexander 2005:79).

NOTES

1. The term *batty bwoy* is one of several derogatory Jamaican colloquialisms for gay men.
2. This negative coverage was a result of the furore which was occasioned by the nomination for Music of Black Origin (MOBO) Awards, a prestigious annual UK music award, of Jamaican dancehall artistes who were alleged to be advocating violence towards gay men in the lyrics of their songs.
3. The pluralisation of the word *condition* (to *conditions*) is expressing the idea that there are many postcolonial experiences.
4. Lacanian refers to the psychoanalytic theorising of French psychiatrist Jacques Lacan, who argued that the unconscious is structured like a language. For more information on Lacan's work, see Jacques Lacan, 'The Mirror Stage as Formative of the Function of the I as Revealed in Psychoanalytic Experience,' in *From Modernism to Postmodernism*, L. Cahoone, ed., pp. 195–199. Oxford: Blackwell, 2003.
5. Saussurean refers to the structuralist theory of Ferdinand de Saussure. Ferdinand de Saussure was a Swiss linguist who has been credited with what was known as the 'linguistic turn' in philosophical understanding. Saussure studied language as a system of signs and suggested that sets of relationships constitute a system of differences which underpin the meaning of particular words. This relational approach to meaning distinguishes Saussure's theory from a traditional positivist methodology which emphasises objectivism and which would perceive each item or word as existing independently of each other rather than in a relational association. Indeed, Saussure's treatment of language rejects such positivist-inspired notions of meaning, seeing language instead as existing within a larger structure or system of relational and differential meanings. For a more extensive discussion of Saussure's theory, see Jonathan Culler, *Saussure*. London: Fontana, 1976, and Jonathan Potter, *Representing Reality: Discourse, Rhetoric and Social Construction*. London: Sage, 1996.
6. JFLAG (Jamaica Association for Lesbians All Sexuals and Gays) was founded in 1998 in a failed attempt to pressure the government to overturn Jamaica's 140-year-old antisodomy law which prohibits sexual acts between men but not women.
7. It would be prudent for me to note that this assertion of masculinity by the Jamaican male is not only demonstrated through homophobia but also

through the demeaning and objectification of women, as evident also in many of the dancehall songs.

REFERENCES

Abrahams-Clivio, T. June 24, 2004. Homosexuality can seem shocking, hard to understand. *The Jamaica Observer.*
Aitkenhead, D. January 5, 2005. Their homophobia is our fault. *The Guardian.*
Alexander, M. J. 2005. *Pedagogies of crossing.* Durham, NC: Duke University Press.
Ashcroft, B., G. Griffiths, and H. Tiffin. 1998. *Key concepts in post-colonial studies.* London: Routledge.
BBC News. 'Timeline: Gay fight for equal rights.' December 6, 2002. Online: www.news.bbc.co.uk
————. UK Edition. September 8, 2004.
Bhabha, H. 1994. *The location of culture.* London: Routledge.
Campbell, S., and G. Tawadros, eds. 2001. *Stuart Hall and Sarat Maharat: Modernity and difference.* London: Institute of International Visual Arts.
Chambers, D. 2003. A discursive analysis of the relationship between heritage and the nation. Unpublished doctoral dissertation, Brunel University, London.
Charmaz, K., and R. G. Mitchell. 1997. The myth of silent authorship: Self, substance and style in ethnographic writing. In *Reflexivity and voice,* R. Hertz, ed. London: Sage.
Clift, S., and S. Forrest. 1999. Gay men and tourism: Destinations and holiday motivations. *Tourism Management* 20 (2):615–25.
Clift, S., and J. Wilkins. 1995. Travel, sexual behaviour and gay men. In *AIDS: Safety, sexuality and risks,* P. Aggleton, G. Hart, and P. Davies, eds. London: Taylor & Francis.
Crichlow, W. E. A. 2004. History, (re) memory, testimony and biomythography: Charting a buller man's Trinidadian past. In *Interrogating Caribbean masculinities,* R. Reddock, ed. Jamaica: University of the West Indies Press.
Culler, J. 1976. *Saussure.* London: Fontana.
David, L. February 5, 2006. Hello sailors: The United Queendom wants you! *Sunday Times.*
d'Hauteserre, A. 2004. Postcolonialism, colonialism and tourism. In *A companion to tourism.* A. Lew, C. M. Hall, and A. Williams, eds. London: Blackwell.
Foucault, M. 1984. The repressive hypothesis. In *The Foucault reader: An introduction to Foucault's thought,* P. Rabinow, ed. London: Penguin.
Gill, R. 1995. Relativism, reflexivity and politics: Interrogating discourse analysis from a feminist perspective. In *Feminism and discourse: Psychological perspectives,* S. Wilkinson and C. Kitzinger, eds. London: Sage.
Hencke, D. October 2, 2004. Holiday firm ends ban on gay couples. *The Guardian.*
Henwood, K., and N. Pidgeon. 1993. Qualitative research and psychological theorising. In *Social research: Philosophy, politics and practice,* M. Hammersely, ed. London: Sage.
Hertz, R. 1997. Introduction. In *Reflexivity and voice,* R. Hertz, ed. London:: Sage.
Hickman, M. March 8, 2006. Changing attitudes: A portrait of gay Britain. *The Independent* (Online edition): www.independent.co.uk.
Holcomb, B., and M. Luongo. 1996. Gay tourism in the United States. *Annals of Tourism Research* 23 (3):711–13.
Hughes, H. 1997. Holidays and homosexual identity. *Tourism Management* 18 (1):3–7.

————. 2002. Gay men's holidays: Identity and inhibitors. In *Gay tourism: Culture, identity and sex.* S. Clift, M. Luongo, and C. Callister, eds. London: Continuum.

Lacan, J. 2003. The mirror stage as formative of the function of the I as revealed in psychoanalytic experience. In *From modernism to postmodernism.* L. Cahoone, ed. Oxford: Blackwell.

Lewis, L. 2003. *Cultures of gender and sexuality in the Caribbean.* Gainsville, FL: University Press of Florida.

Maxwell, J. August 8, 2004. Making justice invisible. *The Jamaica Observer.*

McClintock, A. 1994. The angel of progress: Pitfalls of the term 'post-colonialism.' In *Colonial discourse and post-colonial theory,* P. Williams and L. Chrisman, eds. Harlow, UK: Prentice Hall.

Michalowski, R. J. 1997. Ethnography and anxiety: Fieldwork and reflexivity in the vortex of US–Cuban relations. In *Reflexivity and voice,* R. Hertz, ed. London: Sage.

Mintel International. November 2000. Gay Holidays—UK.

Murray, D. 2007. The civilized homosexual: Travel talk and the project of gay identity. *Sexualities* 10 (1):49–60.

Oppermann, M. 1999. Sex tourism. *Annals of Tourism Research* 26 (2):251–66.

Petridis, A. December 10, 2004. Pride and prejudice. *The Guardian.*

Pinar, W. 2003. 'I am a man': The queer politics of race in cultural studies. *Critical Methodologies* 3 (3).

Potter, J. 1996. *Representing reality: Discourse, rhetoric and social construction* London: Sage.

Pritchard, A., N. Morgan, and D. Sedgely. 1998. Reaching out to the gay tourist: Opportunities and threats in an emerging market segment. *Tourism Management* 19 (3):273–82.

Said, E. 1995. *Orientalism: Western conceptions of the Orient.* 2nd ed. London: Penguin Books.

Stanley-Niaah, S. 2004. Kingston's dancehall: A story of space and celebration. *Space and Culture* 7 (1):102–18.

Stennet, R. August 5, 2002. On gay rights vs. my rights. *Jamaica Gleaner.*

Thwaites, R. September 29, 2000. Gay rights clarification. *Jamaica Gleaner.*

Travel Wire News. November 2, 2005. Las Vegas named one of the top US destinations for gay, lesbian travelers worldwide. Online: www.travelwirenews.com.

————. June 27, 2006. The best in gay and lesbian travel for 2006. Online: www.travelwirenews.com.

Virtue, E. September 17, 2000. Hanging, anti-gay laws to stay—PM. *Jamaica Gleaner.*

Want, P. 2002. Trouble in paradise: Homophobia and resistance to gay tourism. In *Gay tourism: culture, identity and sex,* S. Clift, M. Luongo, and C. Callister, eds. London: Continuum.

Wignall, M. June 17, 2004. Those flamin' homosexuals. *The Jamaica Observer.*

Williams, P., and L. Chrisman. 1994. *Colonial discourse and post-colonial theory.* Harlow, UK: Prentice Hall.

Young, I. M. 2003. The scaling of bodies and the politics of identity. In *From modernism to postmodernism,* L. Cahoone, ed. Oxford: Blackwell Publishing.

Young, R. 2003. *Postcolonialism: A very short introduction.* Oxford:: Oxford University Press.

7 "We Nyammin?"[1]

Food Supply, Authenticity, and the Tourist Experience in Negril, Jamaica

David Dodman and Kevon Rhiney

INTRODUCTION

The act of eating is a process of articulation binding issues of culture, nature, and economy tightly together. Whilst personal food preferences are inherently cultural, the production of food involves not only the organic interaction of soil, seed, and water but also the social and economic relations inherent in agricultural systems. In addition, the consumption of "local" foods is one of the key strategies employed by tourists as they increasingly search for authenticity in their travel experiences.

These linkages have been evident in the context of the Caribbean for centuries. As Mimi Sheller (2003:77) explains, "Contrary to the assumption that it was only the pursuit of gold and other precious metals that drove European exploration, it was as much the desire to acquire new edible, pleasurable, and pharmaceutical substances, things that had direct and powerful effects on the bodies of those empowered to consume them." The relationship between food and tourism is underexplored, yet it is an essential component of the Caribbean tourist attraction, particularly in the context of new tourist interests related to the search for authenticity.

In recent years, there has been increasing interest in the contested nature of food. Much of this is associated with the study of the new geographies of commodities (Watts 1999) and the ways in which food is an important part of material culture (Cook and Harrison 2003; Cook, I. et al. 2004; Duruz 2005). Other work is more explicitly practical in its orientation and addresses the material consequences of particular commodity pathways for agricultural producers in the majority world (Ransom 2001; Robbins 2003; Bryant and Goodman 2004) in relation to new models of "fair trade." In this respect, food is related to broader issues of material culture and economies. As Ho and Nurse (2005) point out for the Caribbean, popular culture is "not just an aesthetic and commercial space where psychic and bodily pleasures are enacted, represented and marketed [but also] an arena where social values and meaning are put on public display, negotiated and contested" (p. xii). The ways in which Jamaican food and concepts of its authenticity are commodified can therefore be seen not only as being relevant to types of place marketing but also to the broader project of understanding the ways in which different cultural texts are circulated and understood by Jamaicans and tourists alike.

In this chapter, we examine the ways in which concepts of authenticity influence the supply and consumption of food by the tourism industry in Negril, Jamaica. Jamaica is one of the Caribbean's leading tourism destinations, with a total of 2.6 million visitor arrivals in 2005, of whom approximately 1.5 million were stopover visitors. The largest proportion (71.6 percent) of these stopover visitors originated from the United States, followed by Europe, Canada, the Caribbean, Latin America, and Japan, respectively. In the same year, gross visitor expenditure was worth US$1,545 million (Planning Institute of Jamaica 2006). Negril is a small town on Jamaica's west coast with a resident population of 5,670 (Statistical Institute of Jamaica 2001). An isolated and remote fishing village until the late 1960s, Negril now has 3,940 hotel rooms (24 percent of Jamaica's total; Jamaica Tourist Board 2002) and is one of Jamaica's main tourist destinations. Despite this growth, Negril has maintained a reputation for being a more laid-back destination than the tourist meccas of Montego Bay and Ocho Rios, with an approach to tourist development epitomised by the local bylaw that buildings must be no higher than the tallest palm tree.

In this chapter, the concept of authenticity is seen to be integrally related to the material aspects of Jamaica's tourism and agriculture industries. After a brief description of each of these, and the interrelationships between them, we discuss the ways in which ideas of authenticity have developed over time and how these are linked in with consumption patterns, particularly the consumption of food. This provides the necessary framework for addressing the ways in which food and authenticity are linked in Negril and assessing the importance of these linkages for both tourism and agriculture in Jamaica. This approach enables us to illustrate the ways in which the production and consumption of food—whether "authentically Jamaican" or not—is integrated into the tourist experience in Negril and to show how the idea of Jamaica as a tourist destination is produced and represented through the production of particular types of food. However, in contrast to much recent work that focuses either on the technical and managerial aspects of tourism (e.g., Jayawardena 2005a, b) or on the experiences of the tourists themselves, this chapter responds to Sheller and Urry's call for a "de-centring of tourist studies away from tourists" (Sheller and Urry 2004:6). In this regard, the production of the tourist experience by local hotels and restaurants is seen as influencing not only tourist experiences but also in (re)producing Jamaican conceptions of identity and self.

FOOD PRODUCTION AND TOURISM IN JAMAICA

Tourism is an important source of income for many developing countries, particularly for small island states such as those in the Caribbean. However, several studies have shown that the recent rapid growth of tourism in many of these islands has failed to alleviate their economic problems (Pattullo 1996; Telfer and Wall 1996; Torres 2003). In many cases, the industry has not been integrated with other sectors of the economy, and there have been high rates of foreign-exchange leakage. If destination countries, particularly those in the Caribbean, are to maximize benefits from tourism development, ways must be found to increase backward economic linkages, including utilizing local food products in the tourism industry. The greatest benefits that can be provided by tourism will occur when the goods and services that are consumed by that sector originate in the host economy (Goodwin 1993).

National governments frequently justified tourism development in small islands based on the assumption that the economic benefits of the industry would filter down to stimulate other sectors in the economy, particularly agriculture, through multiplier effects (Jefferson n.d.; Dahles 1999). Tourist-driven demand for food was expected to stimulate improvements in quality and increases in the quantity of domestic food production (Torres 2003). In addition, it was suggested that sustained growth in tourism would eventually result in the subsequent growth of other industries, including construction and craft production, due to the formation of functional sectoral linkages within the host country over time. However, a number of studies have noted the failure of these inter-sectoral linkages to develop (Bryden 1974; Belisle 1980; Thomas 1988; Torres 2003; Telfer and Wall 1996). A predominant theme in these studies is the tendency for tourism to generate increased food imports which simultaneously damage local agriculture and cause foreign-exchange leakages. Other problems cited between the two sectors include competition for land (Bryden 1974; Belisle 1983); competition for labour resources (Belisle 1980); inflated land values (Belisle 1983); inconsistent food supplies and varied food quality (Belisle 1983); and inflated food prices (Belisle 1983; Thomas 1988).

Food and beverages account for a considerable share of the gross visitor expenditure, especially for stopover tourists. According to the annual travel statistics prepared by the Jamaica Tourist Board (2002), 58.2 percent of tourist expenditure in Jamaica is spent on accommodation inclusive of food and beverages and 7.3 percent is spent on food and beverages alone (Table 7.1). Since the average daily expenditure per tourist for food and beverages only was US$6.70 in 2002 and the total number of hotel-room nights sold was 2.83 million, the total tourist expenditure for food and beverages was on the order of US$18.9 million, although this is probably an underestimate because food and beverages are also included within accommodation expenditures (especially in the case of all-inclusive hotels). This would seem to suggest that food and beverage sales in the tourist sector account for a significant share of the total foreign-exchange earnings for the tourism industry—that is, assuming that the majority of foods consumed are not imported.

Table 7.1   Distribution of Expenditure of Stopover Tourists 2002

	US$ / person /Night	% of Total
Accommodation (including Food and Beverages)	53.39	58.2
Food & Beverages	6.70	7.3
Entertainment	8.72	9.5
Transportation	5.78	6.3
Shopping	10.00	10.9
Miscellaneous	7.15	7.8
Total	91.74	100

Source: Jamaica Tourist Board (2002)

Recent evidence suggests, however, that the majority of the food utilised in the Jamaican tourism industry is imported (Pennicook 2006). Overall, the Jamaican tourism industry experiences a 50 percent foreign-exchange leakage rate (Ramjee Singh 2006). Ramjee Singh (2006) further found that domestic agriculture and country size were the two most important factors in determining the import content for the tourism industry. Some of the general problems cited included paucity of natural resources, a high degree of openness, high vulnerability to natural hazards such as hurricanes, poorly organised domestic farming systems, overt reliance on a few export commodities, and limited ability to exploit economies of scale (Briguglio 1995). These limitations have prevented many states from forming viable backward linkages between their tourism and domestic agriculture sectors.

In addition, the rapid growth in Jamaica's tourism industry has been accompanied by a drastic decline in the agricultural sector. Production levels in the traditional export crops subsector fell by 32.1 percent between the end of the 2003/4 and the 2004/5 crop years, and export earnings declined by 54.7 percent to US$25.7 million over the same period. Jamaica is now a net importer of food, importing US$602 million worth of food whilst exporting only US$193 million worth in 2005 (Planning Institute of Jamaica 2006). The anticipated benefits for the agricultural sector from tourism have therefore failed to materialise, mainly because policymakers have failed to recognise the structural limitations in the relationship between the two sectors. Momsen (1998) argued that the failure of the domestic agricultural sector to forge proper linkages with tourism in the Caribbean is mainly due to the inability of local food producers to meet the demands of many of the islands' hotel industries. This is embedded in traditions that have characterised the Caribbean since slavery, such as agricultural monoculture, which hinders the introduction of alternative crops and limits plans for diversification and innovation. Others have pointed to the deficiencies in marketing, transport, and storage facilities that hinder food delivery to the hotels, restaurants, and retail sectors (Belisle 1980; Torres 2003). In addition to this, the domestic food-production sector is poorly integrated into the local markets, and traditional practices have tended to favour imported food; therefore, the domestic provision of food has rarely been a major objective of Caribbean agriculture (Conway 1993; Jefferson 1972; Beckford 1972).

Yet although several authors have looked at the relationship between the agricultural industry and tourism, many (e.g., Telfer and Wall 1996; Pattullo 1996) have failed to look at the demand-side of the equation resulting from the particular aspects of food consumption by tourists. However, it is increasingly evident that tourist consumption patterns are integrally tied to the way places are marketed. Although Conway (2004:201) suggests that "the globalisation of food consumption habits, and the introduction of more exotic foods and cuisine into the metropolitan cultures of the tourist's home, has led to the hoped-for stimulation of agricultural diversification and production of Caribbean foods," there is little evidence that this has been translated into a quantitative increase in demand for local dishes in Caribbean resorts. A greater understanding of the processes involved in destination marketing and tourist food consumption can therefore help to elucidate the underlying reasons behind the high food import content of many island tourism industries.

Authenticity and the Tourist Experience

Defining the Authentic

It is difficult to define the concept of authenticity and still more complex to identify what is "authentic." Indeed, the *Oxford English Dictionary's* definition of "authentic" as "of undisputed origin" is particularly unsatisfying, as even the origins of material objects are frequently disputed. At one level, therefore, it would be appropriate to say that the quest for authenticity in culture and material culture is inevitably futile. The concept of the authentic is particularly problematic in relation to food. Although particular foods are seen as being intrinsically linked with particular experiences, the actual ingredients or recipes may be far removed from the associated locale. A recent article in the *New York Times* travel section (Passy 2006) discussed the phenomenon of key lime pie in the Florida Keys. Although the limes in the pies are predominantly grown in Mexico, Passy suggests that "A theme still hums through this island chain, audible to every tourist: 'I eat Key Lime pie, therefore I am vacationing in the Keys.' "

However, in relation to the tourist experience, the actual "undisputed origin" of an object is of less interest than the ways in which certain objects and practices are deemed to be "authentic." This quest for authenticity takes on a particular relevance when it is superimposed on the global structures of inequality created by tourists from more-developed countries visiting less-developed nations. As Meethan (2001:90) argues, "The notion of authenticity within tourism analysis is predicated on a false dichotomy between the non-modern, viewed as the authentic, and the modern, viewed as the inauthentic." In addition, various attempts have been made to link the search for authenticity with a widespread sense of dissatisfaction and alienation generated by the modern world, in which (to quote Marx) "all that is solid melts into air." Meethan (2001:91) goes on to suggest that in many cases, the search for authenticity is conducted "to counteract the alienation and loss of modernity." In this case, the search for authenticity revolves around achieving some kind of personal or existential fulfilment.

This search for the authentic increasingly plays an important role in the marketing of tourist attractions. As Timothy and Boyd (2003:239) state, "Tour operators and other travel service providers appear to agree that people seek authentic experiences, as they have started using terms like 'real,' 'authentic' and 'genuine' excessively in their promotional literature." Nevertheless, this promotional literature could contribute to what Ringer (1998) terms the "falsification of touristic landscapes," in which the type of "gaze" (Urry 1990) desired by the target market will come to determine the very nature of the product.

An early, yet subtly nuanced, assessment of authenticity in the tourist experience can be seen in MacCannell's (1973) concept of "staged authenticity." In this version, although "tourists demand authenticity" (p. 600), instead they receive access to the staged authenticity of a "series of special places designed to accommodate tourists and to support their beliefs in the authenticity of their experiences" (p. 589). Many of the distinct spatialities of tourist experiences in the Caribbean are designed—whether deliberately or not—to satisfy the tourists' demand for authenticity, without fundamentally reordering the spatial and social relationships between visitors and tourists. This issue of staged authenticity is evident in at least two ways related to food production and consumption: the types of food that are produced and consumed and the spaces in which the food is served.

Any discussion of authenticity in the Caribbean context must also take into account concepts of creolisation. Creolisation, as the process of creating a distinctly new form of society through the fusion of different cultures, is epitomised in Jamaica's national motto, "Out of Many, One People." In an international context of globalisation, it must also be recognised that Jamaica and the rest of the Caribbean are places where "East and West, North and South, Third World and First World, capitalism and communism, global high tech and local poverty, tourists and drug runners, all collide . . . it becomes increasingly less clear what is inside and what is outside, what is pure and what is impure, what is North and what is South, what is local and what is global" (Sheller 2003:175). Referring specifically to food, the influence of creolisation is clearly recognised in a recently published cookery book, *Eat Caribbean:*

> The distinctive and increasingly popular flavours of island cuisine embrace a fiery fusion of predominantly African, Spanish, English and French cooking styles. Indian, Chinese and Middle Eastern recipes, too, have been adapted and add substantial influence to the mouth-watering melange (Burke 2005: inside front cover).

This demand for authenticity exists in close association with the demand for the exotic. As Ringer (1998:6) points out, one of the main themes in the literature relating to tourism and authenticity has been the "romanticizing of the 'exotic other'." Although some tourist attractions seek to capture particular markets by making the exotic appear familiar, a far more common trend in the present-day Caribbean might be identified as making the familiar appear "exotic." In this perspective, reassuring and nonthreatening activities, foods, or places are packaged in such a way that they satisfy the tourists' demand for difference, without fundamentally challenging their habits or opinions.

The exotic holds a particular thrill in relation to bodily and sensual pleasures. As Wang (1999:362) suggests, "In tourism, sensual pleasures, feelings, and other bodily impulses are to a relatively large extent released and consumed and the bodily desires (for natural amenities, sexual freedom, and spontaneity) are gratified intensively." These bodily desires include the desire to consume exotic foods and drinks. Although this may be related to the search for authenticity, it is defined not by whether it genuinely originates from the destination country but rather by its opposition to that which is experienced in the home context. Certain types of "exotic food" (not to mention exotic cocktails such as piña coladas) are hardly the authentic items consumed by Caribbean residents yet are exotic because of their difference from what is deemed to be the mundane and the everyday.

Consumption, authenticity, and cultural capital

May (1996) explores the consumption of "exotic" food by members of "the new cultural class of young professionals" (p. 57) in the United Kingdom and suggests that "the consumption of such food operates largely as a means of distinction" (p. 57). In this regard, the search for "authentic" or "exotic" food can be seen as an attempt on the part of the tourist to acquire cultural or social capital. According to this model, the most important aspect of the tourist experience is the collection of anecdotes and experiences with which to bolster one's standing amongst one's associates "back home" (Dodman 2004). Mowforth and Munt (1998) suggest that this is most important for the new petite bourgeoisie who—by virtue of their cultural consumption—seek to differentiate themselves from the working classes below (mass-packaged tourists) and the high spending bourgeois middle classes above. This is closely linked with Bourdieu's concept of *habitus,* in which social distinction occurs through the acquisition and consumption of particular cultural objects (Paterson 2006). May (1996) sees consumption as a process through which products become imbued with particular symbolic meanings and suggests that "what we eat has always been a way of proclaiming more about us than how much money we have" (p. 59). Not only does the type of food consumed reflect something about our personalities but also the ways in which this food is acquired—the *how* and *where* of buying and eating. The combination of tourism and food—both of which are important signifiers in postindustrial societies—represents a potent mixture through which individuals can represent themselves to others.

Although people can become world travellers in their own cities through their consumption of different foods from distant places (Duruz 2005), these signifiers become increasingly important when they are combined with the physical process of travel. As Shaw, Agarwal, and Bull (2000) point out, "Tourist practices do not simply entail the purchase of specific goods and services but involve the consumption of signs" (p. 275). The food that is consumed on vacation is superimposed on the type of vacation to define a particular experience that reflects the type of person that the tourist is. The ways in which particular foods and drinks are defined as authentic or exotic, and the ways in which these are seen to grant the consumer certain kinds of cultural capital, can be vividly seen in the following description of the food and restaurant industry in Negril, Jamaica.

Authenticity and Food in Negril

As indicated, the consumption of different foods has long been recognised as an important part of travel. Sheller (2003:81) reports on a dinner described by Thomas Thistlewood in his eighteenth-century Jamaican diary, consisting of "stewed mudfish, and pickled crabs, stewed hog's head, fried liver, etc., quarter of roast pork with pap paw sauce and Irish potatoes, bread, roast yam, and plantains, a boiled pudding (very good), cheese, musk melon, water melon, oranges, French brandy said to be Cognac, punch and porter." Such culinary adventures were perhaps beyond the contemplation of more timid tourists to the tropics throughout much of the twentieth century, yet there is an increasing recognition of the importance of food in the tourist experience. Indeed, food is now marketed alongside other more traditional attractions as not only a pleasant by-product of travel but a reason for visiting new places. A recent report in the British *Sunday Observer* (Hartley 2006) suggests that although "the foodie fever that swept Britain into the European culinary front line has been slow to come to the Caribbean . . . gourmets can now safely graze their way round the islands at tables that compete with London and New York's finest." In the following sections, we identify three ways in which food and authenticity are linked in the particular context of Negril:

the production of "authentic" food; the association of food and place; and the ways in which food is made "accessible" to the tourist (through both spatial and culinary constructions).

Producing authentic food

The first way in which issues of food, tourism, and authenticity are related in Negril is in the ways in which food is purchased and prepared. This section draws on an extensive survey of forty-one hotels and forty-four restaurants in Negril in order to show the ways in which food and tourism and linked and also to illustrate the extent to which the food on offer to tourists represents authentic Jamaican cuisine.

Food-Purchasing Patterns

Hotels and restaurants in Negril obtain their food supplies from several major sources. These sources can be broadly categorised as large wholesalers, small local suppliers, small retailers and supermarkets, and the local produce market located approximately thirty kilometres away in the town of Savanna-la-Mar. Hotels and restaurants may rely on several of these sources to obtain their food needs, and these may be both local and imported. The use of imported food may indicate either a demand for this type of food from tourists or the inefficiency of the local food economy in adequately meeting the industry's demands.

Small local suppliers or higglers are the major suppliers of locally grown food items. Each higgler supplies several hotels and restaurants on a weekly basis. They supply mainly fruits and vegetables, seafood, and a few other products such as eggs. Most of the establishments only have verbal agreements with these higglers, although these relationships may have developed over a number of years. Cooperatives are another form of local food network that provide a significant proportion of local food to hotels and restaurants. This arrangement involves a group of farmers that produce a particular crop or set of crops for a common market. Cooperatives operate on a much more formal basis than higglers. In one particular example, the Sandals all-inclusive chain has created a sophisticated set of relationships with farming cooperatives, including providing these with seeds and some technical support. However, some hotels have complained about the inconsistent supply of high-quality food from cooperatives, a problem that may indicate the limitations of the local food economy, or may be a result of sociocultural factors inhibiting the successful link between the two industries. Large wholesalers are the major suppliers of imported food items to hotels and restaurants in Negril, followed by small retailers and supermarkets. These companies provide a variety of dry, canned, and frozen food products which are supplied to the hotels in bulk.

In an analysis of the food-purchasing patterns of one large restaurant, it is clear that food supplies are purchased from a variety of sources. Vegetables (including tomatoes, cabbages, carrots, and onions) are purchased from a higgler who obtains supplies from the neighbouring parish of St. Elizabeth whilst fruits (including bananas, papaya, and pineapple) come from a variety of other nearby parishes. The case with meat and seafood is more complex: although chicken and lobster are provided by local suppliers; some fish is local and some is imported; and beef and shrimp are imported. However, the situation with vegetables is currently changing, as interviews with suppliers suggest that they are increasingly obtaining and supplying imported vegetables and that the eventual purchasers may not be aware of the provenance of these.

Food-Preparation Patterns

From the data collected in the Negril survey it was clear that a significant amount of the food consumed in the tourism industry is imported. An independent quantitative assessment carried out by the Jamaica Tourist Board in 2005 revealed that while 70 to 90 percent of the food used in traditional small-scale hotels is produced locally, as little as 40 percent of the food consumed in large all-inclusive hotels is grown locally (Pennicook 2006). This may be due to practical concerns: for example, several of the food proprietors in all-inclusive hotels stated that they were able to get higher yields from frozen-vegetable imports than from locally grown vegetables. Since these large all-inclusive hotels entertain a larger guest population (as shown in their higher average room occupancy rates), their focus may be centred more on quantity rather than quality. However, this is particularly important, as all-inclusive hotels—despite their relatively small numbers in Negril—are substantial consumers of food, both by quantity and by value. For instance, Negril contains 29 small hotels (less than 50 rooms) with a total of 769 rooms, and 15 large hotels with a total of 2,609 rooms. These and other similar trends, such as the tendency of larger hotels to specialise in international cuisines and employ overseas and foreign-trained chefs, have far-reaching implications for the local economy in general, especially as it relates to the industry's net retention capacity and the tourist experience in particular in terms of the type of food to which they are exposed (Table 7.2).

Approximately 20 percent of the small hotels surveyed specialise only in Jamaican food, whereas no large hotels specialise only in this. Conversely, a greater proportion of large hotels specialised solely in international dishes. The data therefore suggest that large hotels are more likely to specialise in international cuisine and small hotels in Jamaican cuisine. Large hotels are still substantial consumers of local produce; however, a substantial portion of this is served as meals to staff rather than to guests.

The nationality of the chefs employed in Negril also influences the type of food that is produced. In total, eleven non-Jamaican executive chefs are employed in hotels and restaurants in Negril. Almost all of these chefs are employed in large hotels, and more than half of the large hotels (53 percent) employ non-Jamaican chefs. These chefs are from a variety of countries including the Dominican Republic, Israel, Italy, Japan, Mexico, and Trinidad and Tobago. In addition to the nationality of the chef, the location in which these individuals are trained is also likely to influence the type of food that they produce. Some of the small hotels (21 percent) employed chefs who had received some training overseas, whereas the majority (60 percent) of large hotels employed individuals who had trained overseas. The general trend that can be identified is that large hotels are more likely to employ chefs with foreign nationalities and overseas training, while small hotels predominantly employ Jamaicans chefs with local training.

Associating place and food

The consumption of different foods is juxtaposed with a variety of other aspects of the tourist experience. As Urry (1992:172) describes it, "Tourists experience extremes of heat, taste unexpected dishes, experience heightened passions, hear unusual sounds, encounter new smells, and so on." Distinctive cuisines, therefore, are particular forms of material culture embodying essential qualities of difference and distinctiveness in the tourist experience (Meethan 2001).

Various places, therefore, attempt to illustrate their distinctiveness through associations with particular types of food. In perhaps the most extreme example, this can be seen through promotional material on display at the Negril branch of the multinational fast-food chain Burger King:

> We are proud to offer our customers the best of Jamaica. That's why we use locally grown products on our menu, Content Beef, Best Dressed Chicken and Fish, Blue Mountain Coffee, Fresh Vegetables and Bread from local farmers and manufacturers. By so doing, we are able to support our local agricultural and manufacturing sectors and contribute to the growth of Jamaica's economy.

Tourists themselves make these connections. As "the meanings of commodities are not fixed, but open to reinterpretation at both an individual and

Table 7.2 Characteristics of Food Preparation

	Dish Speciality			Chef Nationality		Mode of Operation	
Hotel Size	Local	Inter- national	Both	Non- Jamaican	Jamaican	All- inclusive	Non All- inclusive
Small Hotels	27	14	59	—	100	8	87
Large Hotels	—	27	73	53	47	92	13

*Percentage of hotels recorded in each category

social level" (Meethan 2001:94–5), the places in which food is consumed are as important as the food itself in determining the nature of the experience. An entry in the guest book at the *Hungry Lion* restaurant (Figure 7.1) by two tourists identified as "DP and DS from Iowa" describes the establishment as having "wonderful atmosphere, terrific food, beautiful people—all things Jamaican in one place" (DP and DS, Iowa, USA, 2005). In its own promotional material, the Hungry Lion restaurant suggests that it "offers an unbeatable fusion of experiences: sumptuous fare in a strikingly beautiful setting, great service, and the best in musical grooves." In this case, the food is just one of many aspects of the overall consumption experience.

Food is also often deemed to acquire positive attributes through its association with a particular place. In the context of Negril, food is often associated with Jamaica or the Caribbean, even when it is not a particularly Jamaican food. This is highly evident in the menu at Jimmy Buffett's Margaritaville on the beach, which includes "Bahama Mama Conch Fritters," "Jamaican Jerk Chilli," and "Ex-Patriated American Chicken" in all

Figure 7.1 The Hungry Lion restaurant.

of which the name of a place is juxtaposed with the description of the food. An entire section of the menu is dedicated to "Yardie-Style" food, described as "Nyammin With These Original Jamaican Flavours." Rick's Café, which was recently featured as one of the *Observer's* best places in the Caribbean "for rum punch at sunset" (Hartley 2006:11), also illustrates this phenomenon. Although "Island Jerk Chicken" is (rightly) described as a "Jamaican tradition," the description of "Hurricane Blackened Tuna" clearly links the climate of the Caribbean (itself invoking exotic yet threatening images of the tropics) with a nonindigenous fish. The "Jamaican Shrimp Alfredo" is ambiguously described as containing "sautéed Caribbean jumbo shrimp," yet neither the linguini pasta nor the Alfredo sauce has any real links to the country or the region.

Increasing accessibility to food

The final mechanism for linking food and authenticity in Negril is related to the production of seemingly authentic foods yet in a way that it is not totally unfamiliar to tourists. This could be seen as the "watering-down" of Jamaican culinary traditions in order to make them seem nonthreatening. However, the search for authenticity does not always require sanitisation, as adventure frequently involves some element of risk. As Urry (1992:181–2) suggests, "Being a tourist appears to involve some striking changes in what is perceived as risky. For example, visitors to an area may be willing to risk illness through, for example, eating contaminated foods (such as local shellfish) or having sexual relations with strangers because of the forms of exotic visual consumption that places such activities in a different context from what is normal and everyday."

One way in which this is evident in Negril is the tactic adopted by the ubiquitous roadside "jerk chicken" vendors (Figure 7.2). Whereas in Kingston and the rest of Jamaica, these vendors prepare their chicken in one spicy style, the vendors in Negril produce two types of chicken—regular "pan" chicken and an alternative "polo" chicken. The "polo" chicken is wrapped in foil and lightly seasoned before cooking over the open fire, resulting in a moister and less spicy meal—which the vendors say is more suited to the tourist palate.

This is taken to an even greater extent in the attempts made by certain all-inclusive hotels to produce authentic Jamaican food. The RIU Tropical Bay resort in Negril is a recently constructed Spanish-owned all-inclusive resort, which offers among its myriad pleasures a "jerk bar" on the beach (Figure 7.3). As well as producing low-cost, locally produced food (in contrast to the generally high reliance on imported food items at all-inclusive hotels), this facility is obviously aimed at providing guests with an "authentic" Jamaican culinary experience. Rather than the traditional Jamaican hard dough, however, the jerk is served with a selection of European breads.

Figure 7.2   Roadside jerk-chicken vendor.

Figure 7.3   The "jerk bar" at the RIU Hotel.

Perhaps this tactic illustrates MacCannell's (1973) concept of "staged authenticity." The ability to purchase food on the street is a response to the tourists' demands for authenticity, yet instead this provides access to a back stage, one of a "series of special places designed to accommodate tourists and to support their beliefs in the authenticity of their experiences" (MacCannell 1973:589). The tourists purchasing chicken from these vendors believe that they are having an "authentic" Jamaican experience, yet are consuming a product specifically designed for the tourist market. This is not to deny that these individuals have a genuine desire to partake in Jamaican culture or cuisine, nor to belittle the ingenuity of the vendors in creating this special product. Rather, it can be seen as a microcosm of the way in which so many tourist-local relationships are mediated through the demands of the tourism industry which hinder the possibility of real individual connections being made.

CONCLUSIONS

It is important to recognise that tourism does not necessarily have a negative impact on cultural authenticity and may in some cases even have a positive effect. For example, Williams (1998) notes that souvenir production may help to sustain craftsmanship and encourage the retention of traditional skills and practices. In a similar vein, it is indeed possible that tourist demands for particular types of food may help to encourage the maintenance of local food traditions in the face of the encroaching globalisation of tastes. However, this depends to a large extent on the ways in which local foods are promoted to visitors and the ways in which the local food network is managed. Instead of acting in competition with each other, there is the potential for mutually beneficial cooperation between the tourism and agriculture sectors. This is of particular relevance in the context of a declining agricultural industry and the search for new means of decreasing the foreign-exchange leakage from the tourism industry.

In this chapter, we have shown how food production and consumption in Negril, Jamaica, are associated with the concept of authenticity. The ways in which food is produced, prepared, and consumed reflect a complicated set of understandings about what is authentically Jamaican and how this is (re)produced in a tourist economy. The association of place and food is seen to play a key role in defining the authentic, whilst the preparation of "exotic" food in a nonthreatening manner speaks to a particular type of "staged authenticity" in this context. These findings all illustrate the ways in which intangible aspects of tourist consumption are becoming increasingly important: As Crick (2002) explains, even friendship with local residents has been commodified in the tourist experience. Yet these ephemeral aspects of tourism also have important material consequences, as an increasing demand for high-quality local food may help to provide sustainable livelihoods for thousands of Jamaican farmers, traders, and merchants.

In recent years, there has been a global shift away from traditional models of mass tourism, and new niche markets are developing rapidly (Dodman 2004). Tourist destinations need to "emphasise alternatives to mass marketing through expanding visitor stay and expenditure" (McElroy 2005:660), and the creation of particular types of culinary experiences—as shown in this chapter—may be a fruitful way of achieving this.

NOTES

1. "Nyammin" is a Jamaican patois term for eating.

REFERENCES

Annual Travel Statistics. 2002. Jamaica Tourist Board. Kingston: Jamaica Tourist Board, Corporate Planning and Research Department.

Beckford, G. 1972. *Persistent poverty: Underdevelopment in plantation economies of the Third World.* New York: Oxford University Press.

Belisle, F. J. 1980. *Hotel Food Supply and local food production in Jamaica: A study in tourism geography.* Unpublished doctoral dissertation. London: University Microfilms International.

———. 1983. Tourism and food production in the Caribbean. *Annals of Tourism Research* 10:497–513.

Briguglio, L. 1995. Small island developing states and their economic vulnerabilities. *World Development* 23 (9):1615–1632.

Bryant, R., and M. Goodman. 2004. Consuming narratives: The political ecology of "alternative" consumption. *Transactions of the Institute of British Geographers* T N.S 29:244–366.

Bryden, J. 1974. *The impact of tourist industries on the agricultural sectors.* Paper read at Proceedings of the Ninth West Indian Agricultural Economics Conference at St. Augustine, Trinidad.

Burke, V. 2005. *Eat Caribbean.* London: Simon & Schuster.

Conway, D. 1993. The new tourism in the Caribbean: Reappraising market segmentation. In *Tourism marketing and management in the Caribbean,* D. Gayle and J. Goodrich, eds. London: Routledge.

———. 2004. Tourism, environmental conservation and management and local agriculture in the eastern Caribbean: Is there an appropriate, sustainable future for them? In *Tourism in the Caribbean: Trends, development, prospects.* D. T. Duval, ed. London: Routledge.

Cook, I. 2004. Follow the thing: Papaya. *Antipode* 36 (4):642–64.

Cook, I., and M. Harrison. 2003. Cross-over food: Re-materializing postcolonial geographies. *Transactions of the Institute of British Geographers N.S* 28:296–317.

Crick, A. 2002. Glad to meet you—my best friend: Relationships in the hospitality industry. *Social and Economic Studies* 51 (1):99–126.

Dahles, H. 1999. Tourism and small entrepreneurs in developing countries: A theoretical perspective. In *Tourism and small entrepreneurs: Development, national policy, and entrepreneurial culture: Indonesian cases,* H. Dahles and K. Bras, eds. New York: Cognizant Communication Offices.

Dodman, D. 2004. Profitability or postmodernity? Changing modes in Jamaican tourism. In *Beyond the blood, the beach and the banana: New perspectives in Caribbean studies,* S. Courtman, ed. Kingston: Ian Randle.

Duruz, J. 2005. Eating at the borders: Culinary journeys. *Environment and Planning D: Society and Space* 23:51–69.

Goodwin, G. 1993. *An Examination of linkages between tourism and the agriculture and manufacturing sectors in the OECS.* St. John's, Antigua and Barbuda: OECS publication.

Hartley, J. March 19, 2006. Bite sized Caribbean. *The Observer* 10–11.

Ho, C., and K. Nurse. 2005. Introduction. In *Globalisation, diaspora and Caribbean popular culture.* Kingston: Ian Randle Publishers.

Jayawardena, C., ed. 2005a. *Caribbean tourism: Visions, missions and challenges.* Kingston: Ian Randle Publishers.

———. 2005b. *Caribbean tourism: People, service and hospitality.* Kingston: Ian Randle Publishers.

Jefferson, O. N.d. Some economic aspects of tourism. In *Caribbean tourism: The economic impact of tourism.* Christ Church, Barbados: Caribbean Tourism Research Centre.

———. 1972. *The Post-War Economic Development of Jamaica.* London: The Gresham Press.

MacCannell, D. 1973. Staged authenticity: Arrangements of social space in tourist settings. *American Journal of Sociology* 79 (3):589–603.

May, J. 1996. A little taste of something more exotic. *Geography* 81:57–64.

McElroy, J. 2005. Managing sustainable tourism in the small island Caribbean. In *The Caribbean economy: A reader,* D. Pantin, ed. Kingston: Ian Randle Publishers.

Meethan, K. 2001. *Tourism in global society: Place, culture, consumption.* London: Palgrave.

Momsen, J. 1998. Caribbean tourism and agriculture: New linkages in the global era? In *Globalization and neoliberalism: The Caribbean context,* T. Klak, ed. pp 115–133. Lanham, Maryland: Rowman and Littlefield.

Mowforth, M., and I. Munt. 1998. *Tourism and sustainability: New tourism in the Third World.* London: Routledge.

Passy, C. March 31, 2006. Searching Margaritaville for the perfect key lime pie, Online: http://travel2.nytimes.com/2006/03/31/travel/escapes.

Paterson, M. 2006. *Consumption and everydaylLife.* London: Routledge.

Pattullo, P. 1996. *Last resorts: The cost of tourism in the Caribbean.* Kingston: Ian Randle Publishers.

Pennicook, P. 2006. The all-inclusive concept: Improving benefits to the Jamaican economy. In *Tourism: The driver of change in the Jamaican economy?,* K. Hall and R. Holding, eds. Kingston: Ian Randle Publishers.

Planning Institute of Jamaica (PIOJ). 2006. *Economic and Social Survey of Jamaica,* 2005. Kingston: Planning Institute of Jamaica.

Ransom, D. T. 2001. *The no-nonsense guide to fair trade.* Oxford: New Internationalist Publications.

Ringer, G. 1998. Destinations: Cultural Landscapes of Tourism. New York: Routledge.

Robbins, P. 2003. *Stolen fruit: the tropical commodities disaster.* London: Zed Books.

Shaw, G., S. Agarwal, and P. Bull. 2000. Tourism consumption and tourist behaviour: A British perspective. *Tourism Geographies* 2 (3):264–89.

Sheller, M. 2003. *Consuming the Caribbean.* London: Routledge.

Sheller, M., and J. Urry. 2004. Introduction: Places to play, places in play. In *Tourism mobilities: Places to play, places in play,* M. Sheller and J. Urry, eds. London: Routledge.

Singh, R. 2006. *Import content of tourism: Explaining differences among small island states. Tourism Analysis,* Volume II, No. 1, pp 33–44.

Statistical Institute of Jamaica. Jamaica Population Census 2001. Kingston: Statistical Institute of Jamaica.

Telfer, D., and G. Wall. 1996. Linkages between tourism and food production. *Annals of Tourism Research* 23 (3):635–53.

Thomas, C. 1988. *The poor and the powerless: Economic policy.* New York: Monthly Review Press.

Timothy, D., and S. Boyd. 2003. *Heritage tourism.* Harlow, UK: Pearson Education.

Torres, R. 2003. Linkages between tourism and agriculture in Mexico. *Annals of Tourism Research* 30 (3):546–66.

Urry, J. 1990. *The tourist gaze.* London: Sage.

———. 1992. The tourist gaze "revisited." *American Behavioural Scientist* 36 (2):172–86.

Wang, N. 1999. Rethinking authenticity in tourism experience. *Annals of Tourism Research* 26 (2):349–70.

Watts, M. 1999. Commodities. In *Introducing human geographies,* P. Cloke, P. Crang, and M. Goodwin, eds. London: Arnold.

Williams, S. 1998. *Tourism geography.* London: Routledge

Part II

Governance

8 Reflections from the Periphery
An Analysis of Small Tourism Businesses within the Sustainability Discourse

Sherma Roberts

. . . Within the current debate on the politics of sustainable tourism, the focus is on ecological and narrowly defined economic processes, rather than the cultural and political framework within which policy choices are made. The limits to growth are both environmental (external) and psycho-social (internal) . . . to ignore these factors is only to undertake partial analysis . . . to fail to contextualize the nature of tourism development (Hall 1994:197–98)

INTRODUCTION

Within traditional economic theory, small businesses are often promoted and encouraged primarily for economic reasons, such as their contribution to socioeconomic development, employment, and gap filling through innovation (Walker and Green,1979; Nafziger 1990; Stanworth and Gray 1991; Teszler 1993). These assumptions have also been credited to small tourism businesses (STBs) (Brohman 1996; Andriotis 2002; Rogerson 2002; Morrison 2002). Added to these primarily macroeconomic assumptions is the hypothesis that STBs also significantly enhance the social and cultural life of the destination (Echtner 1995; Buhalis and Cooper 1998; Main 2001) and are more environmentally appropriate and economically beneficial, particularly when juxtaposed against larger, foreign-owned tourism concerns. It is suggested that these assumptions have been given greater legitimacy within the current discussion on sustainable tourism development, where there is a stress on local ownership, smaller scale, and spreading the benefits within the host community (Brohman 1996; Koh 1996; Dahles 1999, 2000; Rogerson 2002).

The dominance of small tourism businesses within the tourism industry raises the question of whether these entities do indeed possess the capacity to progress destinations towards the goals of sustainable tourism development (STD). Emphasising the environmental dimension of STD, some authors indicate that small tourism businesses, particularly in the accommodation sector, operate from a very tourism-centric definition of sustainable tourism development, which favours economic interests over social and environmental considerations, especially when the economic stability of their businesses is threatened (Berry and Ladkin 1997; Vernon et al. 2003; Revell and Rutherfoord 2003). The reasons advanced by theses authors for STBs' lack of commitment towards the environment included increased competition within the sector, the limited length of the tourism season, and insufficient or no knowledge of what environmental sustainability entails. What has therefore emerged in the literature is that the discourse surrounding STBs and environmentally sustainable practices tends to cluster around economic viability and the internal world of the small tourism business owner without any consideration to the historical, political, and governance structures within which these businesses operate. Taking a political ecology perspective, Gössling (2003) points out that key to an investigation of sustainable tourism development is an understanding of the historical context, local realities, and the interaction of actors that help shape tourism development.

This chapter explores the political context and governance structures within which small tourism businesses are to achieve sustainable tourism development. It takes a more political economy perspective (which I argue is not dissimilar to Gössling's political ecology conceptual framework) and contends that within the current discourse on sustainable (tourism) development the environmental and development discourses have been privileged, with a concomitant subjugation of the political economy or contextual (national) discursive narrative—the latter of which often impede or mediate progress towards sustainability goals. The chapter is mainly conceptual (findings are reported elsewhere) and focuses on small tourism businesses in the Caribbean island of Tobago, which is the smaller of the twin-island Republic of Trinidad and Tobago.

The chapter commences by providing a contextual framework for this discussion by examining the historical, political, and socioeconomic issues that characterize development in Tobago. This discussion is followed by a Foucauldian analysis of discourse and its relevance to understanding the contested nature of sustainable (tourism) development. Consideration is then given to what I regard as the dominant and competing narratives surrounding sustainable (tourism) development, and how these discursive narratives fail to recognize the influence and challenges that the international political economy and indeed the national context place upon small tourism businesses by progress towards sustainable tourism development. Of note is that the discussion centres on *sustainable development* (SD)—the term made popular by the Brundtland Report, also known as *Our Common Future* (World Commission on Environment and Development 1987)—as SD has framed the way in which sustainable *tourism* development has been represented, and is therefore important in providing a broad overview of the debate. Attention is then turned to the exogenous and endogenous challenges that mediate the achievement of sustainability goals for small tourism businesses within a developing country context. The chapter concludes by arguing for a more comprehensive theorizing on sustainable tourism development, particularly given that a lot of what we regard as travel and tourism occurs in peripheral or developing countries.

TOBAGO—THE HISTORICAL, POLITICAL, AND ECONOMIC CONTEXT

Situated 18 miles northeast of Trinidad is the island of Tobago, the smaller of the twin-island republic known as Trinidad and Tobago. The island of Tobago is 116 square miles with an estimated population of 51,000 inhabitants, approximately 4 percent of the national population (1.3M; Policy Research and Development Institute [PRDI] 2000a). Once considered an important economic and geopolitical entity by European imperial powers, the island of Tobago has been transformed over the past three decades from an agrarian economy into what is now mainly a service economy. Specifically, tourism is the major private sector activity in Tobago and, outside of government, is the second largest employer in the island (PRDI 2000b; PRDI and Department of Tourism 2001). However, it is contended that the emerging tourism industry operates within a socioeconomic context of a historical core-periphery *type* of relationship with Trinidad that has impeded the progress of tourism development in Tobago. Lacking the critical infrastructure, human resources, and an adequate enabling environment, which Jenkins and Henry (1982) argue that government should provide for the private sector, progress towards sustainable tourism development objectives has inadvertently become slower and more difficult for Tobago's tourism industry, including small tourism businesses.

Like the other British West Indian colonies, sugar was the mainstay of Tobago's economy until the first half of the nineteenth century, when a combination of factors—including a free labour force, competition from beet sugar, and the removal of protectionist policies—started a downward spiral in the sugar fortunes of the British empire (Ashdown 1979; Brereton 1981). It was in the midst of the island's declining sugar fortunes that the British Crown decided to merge Tobago with Trinidad, making Tobago 'a ward' of Trinidad in 1898 (Ottley 1973). The subordinate position that the union bestowed upon Tobago did not engender strong filial feelings toward Trinidad, and this was exacerbated by the smaller island's lack of political autonomy and economic independence. In postindependent (1962) Trinidad and Tobago, the condition under which the union was forged has led to an ambivalent relationship between the two islands, and many Tobagonians continue to allude to a legacy of neglect, dominance, and inequitable budgetary allocations on the part of the Trinidad-based central government (Weaver 1996). For example, relative to Trinidad, which has a substantial manufacturing and petrochemical sector, Tobago has, since independence, evidenced a higher rate of unemployment and underemployment, lower sectoral diversity, higher poverty levels, lower educational attainment levels, and a higher cost of living (Weaver 1996; Craig-James 2000). Indeed, even in 2003, the current president of the Republic of Trinidad and Tobago found it necessary to allude to the division and inequitable development between the two islands. In his speech, he notes that

> while there is no gain saying that much progress has been made in the development of Tobago, many of the challenges and issues that arose prior to and since the amalgamation of the two islands in 1898 persist. It is almost as if we have come full circle. The objective student of the history of Tobago will conclude that there was not much that has been added to Tobago since annexation (*Tobago News* May 5, 2003).

It is felt that the involuntary, albeit 'necessary,' merger with Trinidad has sowed the seeds for a relationship between the two islands that can at best be described as tense, fractious, and unequal, one that perhaps reflects a form of the internal core-periphery relationship advanced by Andre Gunder Frank (1969). According to Frank, this type of relationship is characterised by uneven development where the core advances, often through exploitative means, at the expense of the periphery. In the case of Trinidad (core) and Tobago (periphery), it is argued that the relationship is not inherently or rather classically exploitative but reflects one of acute imbalance and dependence in terms of decision making, power, policy formulation, and government expenditure (Roberts 1994; Weaver 1996; PRDI 1998b). Elaborating on the tenuousness of Tobago's development and economy, the authors of the Tobago Development Plan state:

> In contrast to Trinidad where government accounts for about 15% of all expenditures, government accounts for 75% of all expenditures in Tobago . . . The Tobago House of Assembly (THA) is currently operating with inadequate recurrent and development budgets that together amount to only 3.1% of the national budget. The budget for 1998 was cut so drastically in comparison with the previous years that it cannot maintain many of the projects previously undertaken to facilitate development of the island. Many of these projects are now at serious risk of failure. Cuts in government expenditure in Tobago amount to direct creation of a slump, business failure and unemployment, and will correspondingly retard the process by which Tobago will end its excessive dependence on the State's budget. (PRDI 1998:9–10)

The foregoing quote alerts us that Tobago's current development is highly dependent upon budgetary allocations from Trinidad—a situation that was contrived a century ago but from which Tobago has been unable to free itself. Consonantly, productive activity and development projects in Tobago, including tourism development, has to be largely financed by disbursements from Trinidad, which is not always dependable or 'equitable.' Indeed, it is suggested that, given Tobago's subordinate or peripheral position within the union of Trinidad and Tobago, the island is placed in a vulnerable position where its progress towards sustainable tourism development is to a large extent dictated (facilitated or limited) by the political (policymaking) and economic machinery in Trinidad.

TOURISM DEVELOPMENT IN TOBAGO

After 1950, the economy of Tobago slowly transitioned from an agrarian base to one dominated by the public sector and tourism. The potential to develop a viable tourism industry in Tobago was seen quite early: In 1957 the colonial legislature established a Hotel Development Corporation to loan money for the construction and expansion of hotels on the island (Brereton 1981). This initiative was followed by the establishment of a tourist board in 1958 and the improvement and extension of the airstrip in Tobago. Where tourism displayed some potential for growth, the industry in Tobago was completely set back by the advent of Hurricane Flora in 1963—approximately one year after the republic gained independence (Roberts 1994). A planning team (set up by the government of Trinidad and Tobago), reporting after Hurricane Flora in 1963, endorsed the previously held notion that tourism development in Tobago should be pursued. Thus, from the early 1980s, one saw a major shift away from plans to develop the agricultural potential of Tobago to a major tourism thrust which included an international airport, a deepwater harbour, increased destination marketing, hotel construction, and several concessions and financial and development incentives to encourage investment in the tourism industry (Budget Speech 1986, 1989, 1990, 1993, 1998). Between 1978 and 1996, Tobago's share of accommodation units increased from about one-third of the national total to over one-half, 90 percent of which are concentrated in the southwestern part of the island (Weaver 1996). However, given the traditional inclusion of Tobago's statistics into national statistics (PRDI 2000a, b), visitor arrival figures for Tobago are only available from 1993. Arrival records reveal that there has been a steady and continuous growth in the number of international tourists visiting Tobago, from 17,433 in 1993 to 86,465 in 2005 (Tourism Development Company 2007).

The growth in arrivals to Tobago is perhaps indicative of a synthesis of well-timed factors, including an increase in funds allocated to strategic and targeted destination marketing, the creation of the Tourism and Industrial Development Company (TIDCO)—an agency designed to 'fast track' the development of tourism in Trinidad and Tobago—a cogent and comprehensive Tourism Master Plan in 1995 that had the commitment of industry and government, and an increase in direct flights to Tobago from America and Europe (Weaver 1996).

Indeed, while tourism plays a limited role in the *national* economy, because of the buoyant oil and gas sector, and related industries operating in Trinidad, its importance in the economic life of Tobago cannot be understated, a situation highly reflective of other tourism-dominated Caribbean islands. Elaborating the importance of tourism to Tobago's economy, Weaver (1996:297) notes that 'while tourism accounted for only 1.4 percent of Trinidad and Tobago's GDP in 1995, due in part to Trinidad's economic diversity, its relative contribution to Tobago's economy has been substantially higher.' More recently, the Tobago Tourism Sector Report (PRDI and Department of Tourism 2001) reveals that the services sector, including tourism, accounts for approximately 98 percent of the Tobago GDP, while manufacturing accounted for less than 1 percent.

The aforementioned figures bring into sharp relief how critically important tourism is to the economy of Tobago. In this regard, it is suggested that the imperative becomes one of balancing the preservation and conservation of the natural and cultural resources, which are the main selling points for Tobago, with making the industry economically profitable—the central thesis of sustainable tourism development. While achieving this balance remains the 'theoretical ideal' of sustainable tourism development, it is propounded that Tobago's tourism development is mediated by a number of factors that ultimately constrain the progress (and the contribution) that the industry, including small tourism businesses, make towards sustainable tourism objectives. It is also asserted that many of these mediating issues are heavily intertwined with the historical and existing relationship between the twin-island state (Tobago House of Assembly, 1999).

Official policy documents have identified some of the critical factors constraining optimal tourism development in Tobago (PRDI 1998; Tobago House of Assembly 1999). Among them are:

- Lack of critical infrastructure
- Training and education
- A weak institutional framework
- Low national priority given to environmental matters

Integral to our understanding of the foregoing constraints is that the Tobago House of Assembly (THA), the body invested with administrative jurisdiction over Tobago's affairs, does not control the foreign exchange earnings of the tourism industry; so most of the abovementioned constraints are dependent upon budgetary disbursement from the central government for improvements. Predictably, these disbursements are subject to cuts, depending on the international price of oil, so that low oil prices in one year can sometimes result in development projects in Tobago being abandoned (PRDI 1998). The fractious and ambivalent relationship that characterises relations between Trinidad and Tobago has also permeated the arena of tourism policy, resulting in an effectively weak institutional framework that retards tourism development in Tobago (Kairi Consultants 2000). Among those responsible for tourism policy and its implementation are the Ministry of Tourism, which is national government; Division of Tourism, which is under the jurisdiction of the Tobago House of Assembly; and TIDCO (recently renamed the Tourism Development Company), which is a parastatal organisation. However, the latter two of these three institutions appear either not to have clearly defined roles or prefer not to function within their respective priority areas, the results of which are ineffectiveness, asymmetrical and unreliable information, and an overall, poorly co-ordinated tourism industry (Second Draft Tourism Policy 1999).

FOUCAULT AND DISCOURSE

The notion of sustainable (tourism) development being a discourse draws upon the thinking of French philosopher Michel Foucault (1999), who argues that knowledge and meaning have been constructed through language and represented in such a way that it acquires widespread acceptance through normalisation techniques (such as education, religion, etc.) and subjects who are authorised to make 'truth' claims (Smart 1985; Hollinshead 1999, Chambers 2003) about a particular topic. For S. Hall (1997:44), 'discourse constructs the topic, [it] influences how ideas are put into practice and used to regulate the conduct of others. Just as discourse "rules in" a certain way of talking about a topic . . . it "rules out," limits and restricts other ways of talking, or of constructing knowledge in relation to a topic.' Elaborating on this, Mills (1997) argues that discourses become sanctioned statements because they have been given some institutionalised force which profoundly influences the way individuals act and think. Sustained discursive formation leads to what Foucault describes as *regimes of truth* (S. Hall 1997; Howarth 2000; Cheong and Miller 2000). In other words, the discourse becomes 'owned' by particular institutions and/or disciplines which make certain 'truth claims' regarding the issue/topic/subject, which becomes accepted as 'truth' and manifests in particular behaviours (Chambers 2003). For Foucault, these institutions are not only enmeshed in the production and construction of knowledge but are also embroiled in relationships of power, which allows them to determine which version of events should be sanctioned and included and which should be excluded (Mills 1997). With reference to tourism development in tropical islands, Gössling (2003:26) states that

> It is important to understand how certain groups of actors can create or shape discourses that lead to the use or conservation of the environment. On a national level, tourism-related discourses usually promote tourism, rendering prominent its advantages. In such cases, discourses are created, maintained and controlled by political and economic elites,

which generally represent the group of actors profiting most from tourism development.

This chapter also argues that our understanding and interpretation of sustainable (tourism) development have been conditioned largely by Western institutions, whose political and economic power have been able to legitimise certain constructions of sustainable development while silencing others (Mowforth and Munt 1998; Carvalho 2001). Thus what has occurred is that sustainable development is defined and talked about ('ruled in') largely in terms of development economics (Holden 2000), where the physical environment is seen as a resource to be exploited, with the view that technology would correct environmental imbalances (Stabler 1997) and environmentalism, where there is acknowledgement that the earth's resources are limited and should be conserved (Redclift 1987).

Carvalho (2001), however, contends that missing from the literature ("ruled out") on sustainable development are critical analyses of how the international political economy impact and challenge the ability of some countries to progress towards sustainable development in a way that is environmentally and socially appropriate. What is also subjugated or "ruled out" from the discourse of sustainable development is the distinctive social, economic and political context in which sustainable development is to be applied and achieved (Wood 1993; Tosun 2001; Gössling 2003). In other words, The Brundtland Report, while theoretically recognising the economic differences between developed and developing countries, still advances principles based upon certain assumptions that do not take into consideration the realities of non-Western countries. To ignore 'context' is to fail to consider the region's broader geographic, historic, economic, and sociocultural characteristics. These constitute an 'external environment' to tourism that effectively determines whether sustainable tourism development can be achieved.

COMPETING DISCOURSES OF SUSTAINABLE DEVELOPMENT

Sustainable development seeks to bridge the gap between development and environmental management. While the notion term *sustainable* has engendered broad support, perhaps because of its 'beguiling simplicity' (O'Riordan 1988), it has also been criticised on the grounds of being too ambiguous, lacking consensus in meaning, and being deceptively simple in its enunciation of what sustainable development requires in practice (Lele 1991; Wood 1993; Burns and Holden 1995; Mowforth and Munt 1998; Butler 1999). So that, while 'many authors and actors involved in the SD debate use the same terminology, there is little in common in substantive terms' (Carvalho 2001:64). But precisely what is at the heart of contention on how sustainable development should be interpreted? Springett (2003) posits that at the crux of the debate are the mechanisms which are utilized by organizations to achieve the aims of SD. That is to say, does sustainable development require that economic development be a prerequisite for environmental management? Or is environmental management and protection a necessary precondition for economic development? Mowforth and Munt (1998:12) suggest that any answer to these questions begets more critical questions such as 'What is being sustained by whom and for whom? Do all interest groups have the same intentions or aspirations in terms of sustainability?'—questions which point to the highly contested and discursive nature of sustainable (tourism) development. As has been intimated above, the discourse of sustainable development has been largely 'owned' by those favouring a more development-economics perspective, with contestation for meaning emerging from environmental organisations and lobbies.

THE DEVELOPMENT-ECONOMICS DISCOURSE

The development-economics (seen as representing a more anthropocentric perspective) interpretation of sustainable development attempts to combine economic objectives with environmentally sound management, in a spirit of solidarity with future generations (Mitlin 1992; Baker et al. 1997). Adams (1993) asserts that this perspective is associated with the reduction of poverty (not inequity) and the utilisation of appropriate technology. In this perspective, resource scarcities are mitigated by investment in new technologies in order to provide technical fixes and substitution mechanisms for environmental problems. For example, new sources of energy might in time replace the nonrenewable energy resources, while the problems of pollution could be dealt with by stringent laws or (less effectively) taxes on polluters (Goodwin 1997). The question that Adams's assertion raises is what exactly is 'appropriate technology'? If indeed it is that technology can solve the problems of environmental degradation associated with high levels of economic growth, then there is no need to reduce economic growth. However, if technology fails to keep pace with the externalities that the market causes, then untold problems lie in wait for future generations (Miller 2001). Hunter (1995) contends that this anthropocentric articulation of sustainable development does not take proper account of the finite nature of natural resources or of basic needs, and if it does, he argues, it is only token.

This particular narrative closely approaches what some authors regard as a weak sustainability position, one in which technological progress and a greater degree of substitution between natural and man-made capital make it possible for sustainability objectives of limits and equity to be achieved (Hunter 1997; Collins 1998; Knowles, El-Mourhabi, and Diamantis 2001). What has therefore emerged in this discourse is a reinforcing of capitalist values but with a 'green front' (Holden 2000; Carvalho 2001; Springett 2003). In other words, the status quo is maintained in favour of capitalist consumption but with an accommodation of environmental practices, to the extent that it benefits commercial interests (Springett 2003). Environmentalists also advance a similar criticism of this interpretation of sustainable development and point to 'the apparent business as usual' aspect of sustainable development, in which economic growth and technological progress proceed in an unfettered fashion. At the very best, they view the report as only tinkering at the margin of a system that has created the current environmental problem.

Carvalho (2001) also stridently criticises the report's position, which gives crucial importance to economic and technological advances in achieving sustainability. He argues that what is omitted from this rationalisation is how the present political economic structures affect the likelihood of some countries adopting development strategies that give the necessary priority to environmental matters. In this vein, Carvalho (2001:63) submits that

> for instance, many attribute ecological problems to poverty and underdevelopment without critically examining the roots of these processes, and tend to propose as solutions mechanisms that reconcile economic growth and environmental protection through technological advances and appropriate development strategies. These assumptions and proposed solutions reflect a degree of northern bias ... Moreover, such solutions are less likely to work in developing areas.

Central to this political economy perspective is that developed countries are regarded as having certain competitive advantages, such as advanced technology, access to credit and cheap inputs, developed infrastructure, and easy access to lucrative markets, all of which combine to create wealth in these countries but which also result in concomitant uneven development in developing countries (Kiely 1995; Mowforth and Munt 1998). Moreover, the process of centralization and concentration of capital in these developed regions are facilitated by certain international institutions—for example, the International Monetary Fund (IMF), the World Bank (WB), and the World Trade Organisation (WTO)—which are accorded enormous power in shaping international development agenda and the economic policy process within developing countries (Wood 1993; Reid 1995; Mowforth and Munt 1998).

The development-economics discourse of sustainable development presents a special issue for developing countries and island economies, particularly given the fragility of ecosystems and the heavy reliance upon the natural resources to meet the needs of the population. It is argued that in island contexts, this interpretation of sustainable development is inadequate for two reasons. First, as previously stated, the technology (and the human resources) may not be present to monitor whether environmental limits are being breached. Second, occluded from this discourse is that economic development in developing countries like Trinidad and Tobago is often mediated by external metropolitan powers and the new globalization movement, which impacts upon overall economic sustainability (Griffin 2002). Compounding the influence of globalization and international economic and political institutions is the legacy of uneven development that characterises the relationship between Trinidad and Tobago.

Small tourism businesses in Tobago therefore operate within a context where their viability is determined to a large extent by global and national actors (Britton 1982; Shaw and Williams 1990, 1998). Having little or no economic and political power to influence decision making at any level, small tourism firms' viability, and hence economic (and by extension environmental) sustainability, appears to be somewhat predetermined. These mediating factors are not taken up in the development-economics discourse of sustainable development. In contrast, progress towards sustainable development and its attendant objectives are seen as universally applicable, irrespective of national context and the inequality present in the international political economy.

THE ENVIRONMENTAL DISCOURSE

Contesting the development economics interpretation of SD presented earlier is the environmental discourse. In the more radical genre of this perspective, nature has intrinsic fundamental rights and values, rather than value ascribed to it by humans (Mitlin 1992). In other words, nature does not need to provide any function or service to humans in order to be of value. The effective outcome of this philosophy would be to keep the use of natural assets to an absolute minimum so as not to deplete intrinsic value more than is absolutely necessary (Hunter 1995). This interpretation of sustainable development is closely aligned to what has been termed the 'strong sustainability' approach, where considerable effort is made to maintain the existing natural resources and environmental stock, most of which are viewed as critical natural capital (Pearce, Barbier, and Markandya 1990; Hunter 1997; Collins 1998).

The main requirement of strong sustainability is that each and every component of the total natural capital that is bequeathed to future generations is at least equal to that inherited. Proponents of strong sustainability argue that natural capital can be substituted *only* by other forms of natural capital. Thus, underpinning this narrow interpretation of intergenerational equity is the 'constant natural assets rule' (passing on of an equivalent stock of natural resources) based on the principles of nonsubstitutability, uncertainty, irreversibility, and equity (Pearce et al. 1990). For nonrenewable resources, this implies minimizing loss for future generations through greater efficiency of use, reuse, and recycling where possible (Goodwin 1997). Likewise, the utilization of renewable resources (such as fresh water, soil, natural ecosystems, etc.) should be restricted to operate within limits imposed by carrying capacity (Reid 1995; Hunter 1997).

The environmental discursive narrative is, however, questioned by some authors (Hughes 1995; Springett 2003) who argue that the language of the debate (sustainable development) has been so owned by the modern economic structures and institutions so that 'even when the environment is brought into economic considerations, the language is borrowed from classical economics: natural *resources,* natural *capital,* stocks of *assets*' (Springett 2003:73). Such semantics help to constrain any vision of re-locating economic activity in society. Mowforth and Munt (1998) regard this environmental discourse of sustainable development as being 'quintessentially Western' in its construction, failing to recognise that for many developing countries 'environmental issues, while important are secondary to the need to enhance economic activity' (Wood 1993:10). Thus, while the developed North can afford to engage various conservation and preservation measures, for many developing countries the focus is on improved social welfare, which is dependent upon economic growth. Hirsch and Warren (1998, cited in Pleumarom, 2002:152) reach a similar conclusion to that of Mowforth and Munt's.

> It is important to note that the concept of 'sustainable development' is deeply rooted in Western environmentalism that often takes the form of 'enlightenment' and is dependent on achieving a certain level of prosperity and development. This however often appears to be at odds with the livelihood-based environmentalism in Southeast Asia and other parts of the Third World, where poor peasants and forest dwellers are struggling to defend and reclaim land and natural resources for economic and cultural survival.

The representation of sustainable development in the environmental discourse not only fails to address social and economic inequalities between and within nations (Mitlin 1992; Hunter 1995; Carvalho 2001) but acts as a hegemonic force in its ability to impose upon non-Western countries environmental practices that proponents of the discourse consider environmentally sustainable. In this regard, (Mowforth and Munt 1998) argue that sustainable development, as a discourse, is a concept charged with power—defined, interpreted, represented, and imposed upon developing countries by those with the power (economic, political) to influence its meaning, namely, First World organisations. Indeed, the environmental discourse raises issues for this discussion in that, while there are compelling reasons for small tourism businesses to safeguard (preserve and conserve) the renewable and non-renewable resources upon which they subsist, it is unlikely that they will be able to conform to this strong sustainability variant of sustainable (tourism) development, particularly given their national context and the economic imperatives that often drive these small tourism firms (Diaz 2001; Revell and Rutherfoord 2003; Vernon et al. 2003). What may therefore be required is an approach to environmental issues which reflects not only the peculiarities of small tourism businesses but small tourism businesses *in* Tobago.

SUSTAINABLE TOURISM DEVELOPMENT

Similar to its antecedent, the notion of sustainable *tourism* development has also become a 'site of contestation,' at the heart of which is the question of what is to be sustained—is it the tourism industry? or is it the destination, including its society, environment, and all economic activity, including the tourism industry? The response to this has resulted in the articulation of two approaches to sustainable tourism development, namely, 'sustainable tourism development,' which posits a multisector approach, and 'sustainable tourism,' which is aligned to a single-sector approach. Hardy and Beeton (2001) assert that 'sustainable tourism,' although it may share some areas of concern with 'sustainable tourism development,' the emphasis on growth in order for business viability to be maintained is indicative of a specific tourism-centric (single sector) agenda. This interpretation of sustainable tourism development closely mirrors the development economics interpretation of SD, in which the focus is on growth, and in this case, the growth of the tourism industry.

Within this discursive narrative, environmental and sociocultural concerns are subjugated or are important to the extent that they assist in the long-term survival of the tourism industry (Butler 1999). In other words, the concerns and demands of the tourism industry must be given primacy over other sectors' wants and desires (Burns and Holden 1995). Moreover, as Hunter (2002) points out, this variant of sustainable tourism development sometimes manifests itself in superficial activities such as beautifying the environment for tourism (clean streets) rather than a radical appraisal of the environmental functioning of a destination area, resort centre, enclave, or, in the case of this research, a small tourism business. Consequently, this highly product-focused, anthropocentric view often leads to relatively little attention being paid to natural resource demands, with environmental side effects of growth only tackled retrospectively if possible and/or economically viable (Stabler 1997). What has thus emerged from this interpretation is a 'greening of business' (Springett 2003) in which issues of exploitation and unfairness remain subjugated and silenced (Springett 2003); and, when they are included in the discourse, it is often as a public-relations exercise geared to ensuring the preservation of the tourism industry rather than to engender genuine participation and equity of resources (Roberts 2000).

Similar to their larger business counterparts, small tourism businesses appear also to act less out of altruism towards the environment and more out of economic interests (Danvers and Long 1996; Revell and Rutherford 2003; Vernon et al. 2003). Drawing on their study of small tourism firms in Caradon, a village in Cornwall, Vernon et al. (2003:60, 64) point out:

Participants in all groups claimed that a range of motives lay behind the actions that they had taken to reduce the environmental impact of their businesses. The most common motive was for financial gain, primarily through a reduction in costs. For most group members, efforts to minimize their use of resources and the volume of waste that they produced were motivated by savings in operating costs: the environmental benefits were a bonus. Where they [small tourism businesses] felt that their financial security was at risk, environmental issues were [again] considered a low priority. Problems such as the limited length of the tourism season, and increased competition within the sector were perceived as significant obstacles to the adoption of environmental practices.

Berry and Ladkin (1997) and Carlsen, Getz, and Ali-Knight (2001) reported similar attitudes towards the environment among SBE owners. These studies point out that small tourism businesses also operate from a very tourism-centric definition of STD. Like the larger businesses, their discourse on sustainability clusters around the weak end of the sustainability matrix, which favours economic interests over social and environmental considerations. Indeed, the economic vulnerability of small tourism firms to internal and external changes makes it even more likely that they would appropriate this discursive narrative in which economic growth is the primary concern

On the other hand, the 'sustainable tourism development' narrative 'acknowledges that tourism is unlikely to be the sole user of resources and that a balance must be found between tourism and other existing and potential activities in the interest of sustainable development' (Wall 1997:45). The language used in this discourse focuses on 'balance' and 'integration,' with its proponents being more closely aligned to the 'environmental front' (France 1999). In this interpretation of STD, the focus is on maintaining and enhancing the quality of life and the quality of the tourist experience at destination areas through the conservation and preservation of the local, natural, built, and cultural resources (Muller 1994; Hunter 1995, 1997). Moreover, within this perspective, inter- and intragenerational equity considerations, as well as the idea of limits, are explicitly addressed (Cronin 1990), the first through community participation initiatives and the encouragement of local ownership and the latter through mechanisms such as carrying capacity, environmental codes of conduct, and sustainability indicators. Emphasis here is on the need for proactive or anticipatory tourism-development planning and the minimization of both renewable and nonrenewable resources through the use of environmental-management instruments and techniques (Stabler 1997; Knowles et al. 2001; Hunter 2002). This interpretation retains an anthropocentric bias but also emphasises the need to protect the natural resources that support social and economic systems rather than the promotion of tourism-related economic growth for its own sake or as an end in itself (Cronin 1990; Hunter 1995, 2002; Swarbrooke 1999; Butler 1999).

Pertinent to this discussion is that this approach to STD, while being more holistic that the single-sector approach, does not engage with the issues of power and politics, neither at the macro- (national and international economic and political structures) nor microlevel (the community; Arnstein, 1969; Midgley et al. 1986; Willis 1995; Joppe 1996) and the implications that these may present for achieving sustainable tourism-development objectives. In this regard, Sharpley (2002:328) asserts that the achievement of 'true 'sustainable tourism development, as posited by Cronin, can only occur if the developmental and environmental consequences of any activity is located within a global socioeconomic, political, and ecological context. This political-economy perspective will be discussed in greater detail later, with specific reference to the international tourism system. Suffice it to say at this point that what Sharpley is arguing is that the political economy of tourism can frequently present a barrier to development in general and STD in particular. Indeed, neither in the foregoing interpretation of the sustainable development of tourism nor in this discursive narrative is this political-economy approach privileged. In fact, I argue that the present status quo (political and economic) has been largely taken for granted by these two interpretations of sustainable tourism development—a feature mirrored in its parental paradigm. It is contended that this balanced interpretation of sustainable tourism development only tinkers on the margins of environmentalism and fails to challenge the very foundations that underpin current discourses. Gössling (2003:16) notes that '. . . the need to "make tourism sustainable" is usually emphasised, but it is generally not questioned if the fundamental problems underlying sustainable tourism development can be resolved.'

SMALL TOURISM ENTERPRISES AND THE MULTISECTOR APPROACH

So what of small tourism firms within this extra-parochial, multi-sector, more balanced interpretation of sustainable tourism development? A review of the literature has highlighted two very interesting patterns. First of all is that current research on small tourism businesses and sustainability have failed to articulate an understanding of or an appreciation for this multisector approach, so research in this area has tended to focus either on the environmental dimension of STD or the economic dimension (Danvers and Long 1996; Horobin and Long 1996; Dahles 1997, 1999; van der Duim 1997; Bras 1997; Berry and Ladkin 1997; Gartner 2001; Revell and Rutherfoord 2003; Vernon et al. 2003). Moreover, scholars writing about STBs and sustainability issues within the framework of a developing country tend to privilege the economic dimension of sustainability when referring to small tourism businesses (Dahles 1997, 1999; van der Duim 1997; Bras 1997; Gartner 2001). Concomitantly, there is a veritable silence on environmental and sociocultural sustainability with relevance to STEs, the exception to this being Tosun (2001). In contrast, scholars writing about small tourism businesses and sustainability issues in the context of a developed country primarily focus on the environmental dimension (Danvers and Long 1996; Horobin and Long 1996; Berry and Ladkin 1997; Revell and Rutherfoord 2003; Vernon et al. 2003).

A reading of current research in the area suggests that within this perspective most authors seem to equate STD with the physical/natural environment. For example, after engaging with the idea that sustainable (tourism) development is a contested concept, vague and open to interpretation, Vernon et al. (2003:52), writing about STBs in East Cornwall, go on to state that 'only a small number of studies have attempted to examine the responses of micro-businesses to environmental sustainability.' Similarly, Carlsen et al. (2001) refer to sustainable (tourism) development almost as if the physical environment is the only dimension of STD. Consonantly, there is a marked reticence or very oblique reference to the social and cultural dimensions of STD with reference to small tourism businesses. Indeed, as Wood (1993:12) notes:

> Our Common Future suggested that a principle of sustainable development should be the merging of economic, environmental and social considerations into an integrated framework to guide human activity. In reality however the tendency has been to focus activity on one component, dependent to a large extent on the proponents' world view. For example Third World governments have continued to concentrate on issues of economic development based on Western modernization. . . . Conversely the focus in the North has been on the environment, where concern for specific species preservation . . . have often been characterized by an emphasis on protection through exclusion.

Thus, it is suggested that current research on small tourism businesses have appropriated the a priori notions of sustainability, wherein economic and environmental rationality are privileged above social and cultural concerns. Moreover, developing countries' perspectives tend to authorise the discursive narrative of sustainability and economic imperatives, while developed countries' perspectives on STBs and sustainable tourism development privilege a more environmental discourse. Consequently, the more balanced, multisector approach to the sustainable development of tourism is hardly ever considered in the analyses of small tourism businesses and sustainable tourism development.

POLITICAL ECONOMY PERSPECTIVE—THE INTERNATIONAL TOURISM SYSTEM AND SMALL TOURISM ENTERPRISES

Drawing on the Frankian (1969) perspective of development, which posits that capitalist development in the core continuously creates and perpetuates underdevelopment in the periphery, Britton (1996) suggests that developing countries within the tourism industry are at a disadvantage by virtue of the structural organisation of international tourism. In his conceptualisation, the international tourism system comprises a hierarchy where at the apex are those tourist companies that have their headquarters in the metropolitan market countries and that dominate international tourism. At the intermediate level are the branch offices and local associates of these transnational companies, which organize most of the tourism in developing countries. And at the base lie those small-scale tourism enterprises of the destination that are marginal but dependent upon the transnational companies. The ultimate consequence of this hierarchical structure is that while all participants in the industry profit to a degree, the overall direction of capital accumulation is up the hierarchy (Britton 1996). Indeed, the literature suggests that much of tourism development in developing countries is driven by international tour operators who have the power to decide whether or not a specific destination is to be favoured (Tosun 2001). The role that small tourism enterprises are assigned in this system is rather limited as 'external' forces largely dominate their success and/or failure (Dahles 1999). This point is critical because Britton's analysis suggests that the sustainability contributions of small tourism businesses are to some extent controlled by a system over which they have no control. Indeed, Brown (1998) asserts that the tendency towards oligopolistic control and vertical integration in the international tourist system mean that small, locally owned tourism businesses have little chance of surviving. Moreover, the economic power that large companies wield often means that successful small tourism businesses often face being taken over. For example, Cater (1996) notes that the Belizean ecotourism initiatives, which started off as being small, locally owned concerns, now have 65 percent expatriate involvement.

Similarly, an UNCTAD United Nations Conference on Trade and Development report on the 'Sustainability of International Tourism in Developing Countries' (Diaz 2001) found that the economic, social and environmental sustainability of tourism in Third World countries are threatened by 'high levels of financial leakages, escalating predatory practices and anti-competitive behaviour' of travel and tourism corporations based mainly in Europe and the United States. The UNCTAD report further points out that the combined impact of these factors undermines the economic viability of local enterprises and the ability of countries to allocate necessary resource for environmental protection and sustainable development (Diaz 2001). Relatedly, Revell and Rutherfoord (2003) assert that an examination of the governance structures must inform our understanding of the environmental practices (or lack thereof) of small firms. They succinctly posit that

> A persistent theme running through much of the existing literature is that small businesses lack the characteristics that would otherwise enable them to engage effectively with the 'sustainable development' agenda . . . we argue that there are wider issues to consider. Governance

structures and policy arrangements play an equally important part in influencing the environmental practices of small firms. There is therefore a need to focus not just on the internal world of owner-managers and their reasons for responding in a certain way towards environmental pressures but also on *the contextual, structural factors that have shaped their experience of reality.* (author's emphasis; Revell and Rutherfoord 2003:27)

The point made by Revell and Rutherfoord very much reflects the position of this current discussion. By placing emphasis on the nature of political and economic structures in relation to sustainable tourism development, the political-economy approach raises the profile of notions of power that are enmeshed within the discourse on sustainable (tourism) development. What is revealed is that small tourism businesses have little power to shape or change the present international tourism system; so their contributions to particular dimensions of sustainability are controlled by those with the power (economic and political) to determine

Figure 8.1   Governance structures mediating the discourse(s) of sustainable tourism development.

the discourse (knowledge and praxis). Figure 8.1 provides a conceptual framework of this discussion.

While the international tourism system has the ability to affect the sustainability contributions of small tourism firms, it is also acknowledged that endogenous national factors can also mediate the contribution that small tourism firms make to sustainable tourism development.

ENDOGENOUS CHALLENGES TO SUSTAINABLE TOURISM DEVELOPMENT

As the Tobago case study suggests, the question of sustainable tourism development in developing countries is often mediated by a number of realities including underdevelopment, historical antecedents, and lack of/low priority environmental awareness and practice. A summary examination of the literature indicates that tourism development in developing countries is often promoted on the basis that it can help to redress the problems of high levels of unemployment, low per capita income, a declining agricultural sector, few or no manufacturing industries, and, consequently, high levels of foreign debt (Jenkins and Henry 1982; Burns 1993; Dieke 1993; Oppermann and Chon 1997; Din 1997; Freitag 1994). Even with the 'new call' for tourism development founded on the principles of SD, this economic rationale continues to dictate tourism plans and policies of developing countries (Pantin 1999; Tosun 2001). For example, Tosun (2001) notes that, despite the inclusion of a sustainable tourism development agenda in tourism planning in Turkey, it is the priorities of the national economy—such as debt repayment and foreign currency earnings—that dictate the type of tourism practice that is pursued.

For many developing countries, then, economic imperatives have mitigated or even obscured their ability to pursue the long-term goals integral to sustainable tourism development. Concomitant to this economic emphasis is that the underlying policy arrangements in many developing countries often favour larger, often foreign-owned tourism businesses, as they are perceived to be greater income and employment generators (Britton 1996; Dahles 1999). This point is borne out by Dahles (1997, 1999), who notes that government policy in Indonesia is often favourable to large tourism concerns while regarding small (tourism) businesses as obstacles to economic/tourism development. Van der Duim (1997) reports similar findings based upon his research in Costa Rica. Thus, it is argued that developing countries, by adhering to a very weak variant of weak sustainability, present a formidable challenge for small tourism businesses to progress towards and meet sustainability objectives.

Indeed, where the economic imperative is high, environmental matters are often given low priority (Tosun 2001; Sasidharan et al. 2002; Weaver 2002; Pleumarom 2002). This is not a desirable situation, especially since it is the natural, built, and human environments that most of these countries employ as 'pull factors.' Relatedly, Sasidharan et al. (2002) argue that because tourism development in developing countries has the potential to produce negative environmental impacts, thereby altering the ecological resources of host destinations, governments of these countries should be giving environmental matters greater priority. There is therefore an argument for the state to play a directive role in facilitating environmentally favourable structural change (Töpfer 1989 cited in Baylis, Connell, and Flynn, 1998). For many countries, however, it is not that legislation governing the environment is nonexistent but that it is not enforced. Moreover, environment plans and impact assessments are quite costly to conduct and, when juxtaposed against other seemingly more pressing issues such as the provision of water and electricity, invariably take a lower priority. The lack of political will and cost consideration are compounded by human resource deficiencies (Tosun 2001; Wood 1993).

Faced with a broad range of socioeconomic problems, countries of the developing world continue to place emphasis on economic development, often at the expense of the environment. Wood (1993:16) posits that this is further complicated by developing countries' suspicion of the environmental agenda of the North, viewing it as 'form of ecological imperialism.' Thus, while it may be likely that, like their counterparts in the North, small tourism enterprises in developing countries understand the virtue and value of preserving and conserving the natural resources upon which many of them exist, they may not have the power to influence the broader political and economic environment in which they operate, particularly given the development priorities of many of these developing countries.

The discussion so far indicates that issues of inequity, imbalance, and imperatives superintend the actions and policies of developing countries. The intention of the foregoing was to present the realities of developing countries particularly within the context of sustainable tourism development. Emerging out of the discussion is that small tourism businesses operating within the context of a developing country may find it quite onerous to deliver the goals of intra- and intergenerational equity and to embrace the ideas and practices of resource conservation as enunciated by the Brundtland Report. This finding raises the question as to the applicability of the current interpretations of STD. It is suggested that given the location of Tobago within the global and national arenas, that greater utility may be found in adopting a more flexible interpretation. Indeed, if sustainable (tourism) development is regarded as discourse in which power/knowledge is capillary and insidious (Hollinshead 1999; Chambers 2003), then it would not be illogical to advance an interpretation of STD that challenges the dominant narratives. Some authors obliquely refer to this approach as flexible, intimating that sustainable (tourism) development should allow for greater flexibility in its interpretation, given that alternative interpretations are inevitable and that different countries and regions have different concerns (Wood 1993; Hunter 1997, 2002; Griffin 2002; Gössling 2003). The following considers the utility and implications of a flexible interpretation of sustainable tourism development.

The Utility of a Flexible Approach to Sustainable Tourism Development

It is suggested that a flexible approach to sustainable tourism development should reflect the differing nature and extent of social, economic, and environmental problems but still be underpinned by the broader goals of sustainable development. As intimated earlier, such a position specifically acknowledges the differences between developed and developing countries and their attempts to meet STD objectives. For Schmid-heiny (1992, cited in Wood, 1993:8), 'A flexible approach to sustainable development is particularly important in the developing world, given the varying capabilities of such nations to address the issue.' A similar perspective is shared by Griffin (2002:30) as he observes that 'It is far more difficult to be optimistic about the prospects of tourism developing sustainably in less developed nations . . . it may be that we in the developed world have to allow those nations more latitude with their interpretation of sustainability.

For this author, a flexible approach to sustainable tourism development does not mean a total rejection of Western theorizing on this topic. Rather, I feel that the flexible approach provides a space to centre our concerns and worldviews as it pertains to the discourse on sustainable tourism development—a view that takes into consideration our reality. Not to engage in critical investigation of this topic is to render our realities invisible. Some may argue that while there seems to be a logical and practical usefulness in evolving a more flexible approach to sustainable tourism development, academic scepticism requires that one asks such questions as 'Does a flexible definition of sustainable development take us deeper into the labyrinth of ambiguity, vacuous rhetoric, and competing discursive formations?' Does it, as Sharpley (2000:1) contends, 'neatly side-step the need for a concise definition'? The answer to these questions could indeed be affirmative, resulting in definitional anarchism, particularly in cases where little or no justification is given for the interpretation that is adopted. However, the view is taken here that the discursive nature of STD necessarily lends itself to new and competing interpretations that recognise contextual differences as well as inadequacies in present interpretations. That said, it is felt that any discourse of sustainable tourism development that is appropriated should be arrived at through what Hunter (2002:12) calls a 'self-conscious process,' which provides a reasonable and defensible justification for a particular interpretation.

Conclusion

The tourism literature has treated the notion of sustainable tourism development as an umbrella term that can be adopted by *all* destinations and entities *in the same way*. This, however, should not be surprising, since the drafters of the report, while recognising the problems in developing countries, never clearly elucidated the challenges that developing countries face beyond the need for sustained economic growth and technological advancements. Articulated also is that the international political and economic institutions (e.g., the World Bank and the International Monetary Fund) have a role to play in facilitating the progress of these countries towards sustainable development (WCED 1987). With the exception of a few scholars (Mowforth and Munt 1998; Pantin 1999; Tosun 2001; Diaz 2001; Sasidharan et al. 2002; Gössling 2003) research on sustainable tourism development has also ostensibly mirrored the Brundtland Report by failing to ask pertinent questions relating to the achievement of STD in developing countries. The chapter has also suggested that the challenges that developing countries face in their progress towards sustainable tourism development are different from that of developed nations, and it is important that this be acknowledged in assessing the assumptions surrounding small tourism businesses and STD. In particular, research on the island of Tobago has shown that the achievement of sustainable tourism development objectives for STBs can be difficult if one's historical and, by extension, political and financial legacies are that of dependence and subordination (Roberts, 2004. It is these contextual issues that are often excluded from the dominant sustainability narratives but which, I have argued, are critical to our interrogation and operationalisation of the noble objectives embedded in sustainable development.

New perspectives in Caribbean tourism call for a certain degree of 'decolonization,' of resistance to what is often the perceived view. As sustainable tourism development gains greater currency in terms of our understanding, it is important that equal consideration be given to how it is interpreted, understood, and operationalised in peripheral destinations like Tobago and by small tourism businesses. Critically, all development processes are political, so any discussion of development must engage the supranational and national political actors that have the power to shape development. Sustainable tourism development, as a 'new' development agenda, must of necessity consider and privilege this political economy and national discourse as part of its interpretation and representation. This chapter has called for the discourse to be broadened so that the realities of 'the Other' become privileged. This broadening of the discourse allows for thinkers, politicians, and technocrats in developing countries to interrogate the construction and meaning of STD and reshape it in a way that has relevance for achieving improved sustainable businesses and livelihoods.

REFERENCES

Adams, W. M. 1993. *Green development—environment and sustainability in the Third World*. London: Routledge.

Andriotis, K. 2002. Scale of hospitality firms and local economic development—evidence from Crete. *Tourism Management* 23 (4) 333–341.

Arnstein, S. 1969. A ladder of citizen participation. *Journal of American Institute of Planning* 35, 216–24.

Ashdown, P. 1979. *Caribbean history in maps.* Port-of-Spain, Trinidad: Longman.

Baker et al. 1997. *The politics of sustainable development.* London: Routledge.

Barclay, L. S. 1993. Black owned micro-enterprises in Trinidad and Tobago. In S. Ryan, and L. Barclay, eds., *Sharks and sardines: Blacks in business in Trinidad and Tobago.* Port-of-Spain, Trinidad and Tobago: ISER, University of the West Indies.

Baylis, R., L. Connell, and A. Flynn. 1998. Company size, environmental regulation and ecological modernization: Further analysis at the level of the firm. *Business Strategy and the Environment* 7 (5):285–96.

Berry, S., and A. Ladkin. 1997. Sustainable tourism: A regional perspective. *Tourism Management* 18 (7) 430–40.

Bras, K. 1997. Small-scale entrepreneurs: Strategies of local guides on the island of Lombok, Indonesia. In H. Dahles, ed., *Tourism, small entrepreneurs and sustainable development—case studies from developing countries.* Tilburg, Netherlands: ATLAS.

Brereton, B. 1981. *A history of modern Trinidad.* London: Heinemann.

Britton, S. G. 1982. The political economy of tourism in the Third World. *Annals of Tourism Research* 9:331–58.

Britton, S. G. 1996. Tourism dependency and development: A mode of analysis. In Y. Apostolopolus et al., eds., *The sociology of tourism—theoretical and empirical investigations.* London: Routledge.

Brohman, J. 1996. New directions in tourism for Third World development. *Development and Change* 23 (1):48–70.

Brown, O. 1998. In search of an appropriate form of tourism for Africa: Lessons from the past and suggestions for the future. *Tourism Management* 19 (3):237–45.

Budget Speech. 1981. Government of Trinidad and Tobago, December 15, 1980.

———. 1986. Government of Trinidad and Tobago, December 17, 1985.

———. 1989. Government of Trinidad and Tobago. December 16, 1988.

———. 1990. Government of Trinidad and Tobago, December 22, 1989.

———. 1993. Government of Trinidad and Tobago, November 20, 1992.

Buhalis, D., and C. Cooper. 1998. Competition or co-operation? Small and medium sized tourism enterprises at the destinations. In E. Laws, ed., *Embracing and managing change in tourism—international case studies.* London: Routledge.

Burns, P. 1993. Sustaining tourism employment. *Journal of Sustainable Tourism* 1 (2):81–96.

Burns, P., and A. Holden. 1995. *Tourism: A new perspective.* London: Prentice Hall.

Butler, R. 1999. Sustainable tourism: A state-of-the-art review. *Tourism Geographies* 1 (1):7–25.

Butler, R. W. 1998. Sustainable tourism—looking backwards in order to progress? In M. Hall, and A. Lew, eds., *Sustainable tourism: A geographical perspective.* Essex, UK: Longman.

Carlsen, J., D. Getz, and J. Ali-Knight. 2001. The environmental attitudes and practices of family businesses in the rural tourism and hospitality sectors. *Journal of Sustainable Tourism* 9 (4):281–97.

Carvalho, G. 2001. Sustainable development: Is it achievable within the existing international political economy context? *Sustainable Development* 1 (1):61–73.

Cater, E. 1996. Ecotourism in the Caribbean: A sustainable option for Belize and Dominica? In L. Briguglio et al., eds., *Sustainable tourism in islands and states: Case studies.* New York: Pinter.

Chambers, D. 2003. *A discursive analysis of the relationship between heritage and the nation.* Unpublished doctoral dissertation, Brunel University, UK.

Cheong, S. M., and M. Miller. 2000. Power and tourism: A Foucauldian observation. *Annals of Tourism Research* 27 (2):371–90.

Collins, A. 1998. Tourism development and natural capital. *Annals of Tourism Research* 26 (1):98–109.

Craig-James, S. 2000. *The social and economic history of Tobago (1838–1990): Persistent poverty in the absence of adequate development strategy.* Tobago, Scarborough: PRDI, Office of the Chief Secretary, THA.

Cronin, L. 1990. A strategy for tourism and sustainable development. *World Leisure and Recreation Research* 32 (3):12–18.

Dahles, H. 1997. Tourism, petty entrepreneurs and sustainable development. In H. Dahles, ed., *Tourism, small entrepreneurs and sustainable development—case studies from developing countries.* Tilburg, Netherlands: ATLAS.

———. 1999. Tourism and small entrepreneurs in developing countries: A theoretical perspective. In H. Dahles and K. Bras, eds., *Tourism and small entrepreneurs, development, national policy and entrepreneurial culture: Indonesian cases.* New York, New York: Cognizant Communications.

———. 2000. Tourism, small entrepreneurs and community development. In G. Richardsand D. Hall, eds., *Tourism and sustainable community development.* London: Routledge.

Danvers, H., and J. Long. 1996. All in the mind? The attitudes of small tourism businesses to sustainability. In *Issues relating to small business in the hospitality and tourism industries: Proceedings of a conference, Leeds, Spring, 1996.* Leeds, UK: Leeds Metropolitan University.

Diaz, D. 2001. *The viability of sustainability of international tourism in developing countries.* Symposium on Tourism Service, February 22–23. Geneva: World Trade Organisation.

Dieke, P. 1993. Tourism and development policy in the Gambia. *Annals of Tourism Research* 20:423–49.

Din, K. 1997. Indigenization of tourism development: Some constraints and possibilities. In M. Oppermann, ed., *Pacific Rim tourism.* Oxon, UK: CAB International.

Echtner, C. 1995. Entrepreneurial training in developing countries. *Annals of Tourism Research* 22 (1):119–34.

Foucault, M. 1999. The incitement to discourse. In A. Jaworski and N. Coupland, eds., *The discourse reader* (pp. 514–22). London: Routledge.

France, L. 1999. *The Earthscan reader in sustainable tourism.* London: Earthscan.

Frank, A. G. 1969. The development of underdevelopment. *Monthly Review* 18 (4):17–31.

Freitag, T. 1994. Enclave tourism development: For whom the benefits roll? *Annals of Tourism Research* l. 21 (3):538–54.

Gartner, W. 2001. Issues of sustainable development in a developing country context. In S. Wahab and C. Cooper, eds., *Tourism in the age of globalisation.* London: Routledge.

Goodwin, B. 1997. *Using political ideas,* 4th ed. Chichester, UK: John Wiley & Sons.

Gössling, S. 2003. *Tourism and development in tropical islands: Political ecology perspectives.* Cheltenham, UK: Edward Elgar.

Griffin, T. 2002. An optimistic perspective on tourism's sustainability. In R. Harris et al., eds., *Sustainable tourism: A global perspective.* Oxford: Butterworth-Heinemann.

Hall, C. M. 1997. *Tourism and politics: Policy, power and place.* Chichester, UK: John Wiley & Sons.

Hall, S. 1994. The work of representation. In S. Hall, ed., *Representation: Cultural representation and signifying practices*. London: Sage.

Hardy, A. L., and R. Beeton. 2001. Sustainable tourism or maintainable tourism: Managing resources for more than average outcomes. *Journal of Sustainable Tourism* 9 (3):169–92.

Holden, A. 2000. *Environment and tourism*. London: Routledge.

Hollinshead, K. 1999. Surveillance of the worlds of tourism: Foucault and the eye-of-power. *Tourism Management* (20):7–23.

Horobin, H., and V. Long. 1996. Sustainable tourism: The role of the small firm. *International Journal of Contemporary Hospitality Management* 8 (5):15–19.

Howarth, D. 2000. *Discourse*. Buckingham, UK: Open University.

Hughes, G. 1995. *The Cultural Construction of Sustainable Tourism*. Tourism Management 16, 49–59.

Hunter, C. 1995. On the need to re-conceptualise sustainable tourism development. *Journal of Sustainable Tourism* 3, 155–65.

———. 1997. Sustainable tourism as an adaptive paradigm. *Annals of Tourism Research* 24 (4):850–67.

———. 2002. Aspects of the sustainable tourism debate from a natural resource perspective. In R. Harris et al., eds., *Sustainable tourism: A global perspective*. Oxford: Butterworth-Heinemann.

Jenkins, C., and B. Henry. 1982. Government involvement in tourism in developing countries. *Annals of Tourism Research* 9, 499–521.

Joppe, M. 1996. Sustainable community development revisited. *Tourism Management* 17 (7):475–79. Kairi Consultants, Port-of-Spain.

Kairi Consultants. 2000. Final report—preparation of rolling three-year tourism plan. Trinidad and Tobago.

Kiely, R. 1995. *Sociology and development—the impasse and beyond*. London: UCL Press.

Knowles, T., J. El-Mourhabi, and D. Diamantis. 2001. *The globalization of tourism and hospitality: A strategic perspective*. London: Continuum.

Koh, K. 1996. The tourism entrepreneurial process: A conceptualization and implications for research and development. *The Tourist Review* 51 (4):24–41.

Lélé, S. 1991. Sustainable development: A critical review. *World Development* 19 (6):607–21.

Main, H. C. 2001. The use of Internet by hotels in Wales: A longitudinal study from 1994 to 2000 of small and medium enterprises in a peripheral location with a focus on net technology. In P. Sheldon et al., eds., *Information and communication technologies in tourism 2001*. New York: Springer-Verlag Wien.

Midgley, J., et al. 1986. *Community participation, social development and the state*. London: Methuen.

Miller, G. 2001. The development of indicators for the promotion of sustainable tourism. Unpublished doctoral thesis, University of Surrey, UK.

Mills, S. 1997. *Discourse*. London: Routledge.

Mitlin, D. 1992. Sustainable development: A guide to the literature. *Environment and Urbanization* 4 (1):111–24.

Morrison, A. 2002. Small hospitality businesses: Enduring or endangered? *Journal of Hospitality and Tourism Management* 9 (1):1–11.

Mowforth, M., and I. Munt. 1998. *Tourism and sustainability—new tourism in the Third World*. London: Routledge.

Muller, H. 1994. The thorny path to sustainable tourism development. *Journal of Sustainable Tourism* 2 (3):131–36.

Nafziger, E. W. 1990. *The economics of developing countries*, 2nd ed. Upper Saddle River, NJ: Prentice-Hall.

Oppermann, M., and K. Chon. 1997. *Tourism in developing countries.* London: Thomson.

O'Riordan, T. 1988. The politics of sustainability. In R. K. Turner, ed., *Sustainable environmental management: Principles and practice.* London: Belhaven.

Ottley, C. R. 1973. *The story of Tobago: Robinson Crusoe's island in the Caribbean.* Longman: Port-of-Spain, Trinidad.

Pantin, D. 1999. The challenge of sustainable development in small developing states: A case study on tourism in the Caribbean. *Natural Resources Forum* 23, 221–33.

Pearce, D., E. Barbier, and A. Markandya. 1990. *Sustainable development—economics and environment in the Third World.* Aldershot, UK: Edward Elgan,

Pleumaron, A. 2002. How sustainable is Mekong tourism? In R. Harris et al., eds., *Sustainable tourism: A global perspective.* Oxford: Butterworth-Heinemann,

Policy Research and Development Institute. 1998. Tobago Development Plan, Report No. 2—Medium-Term Policy Framework of Tobago, 1998–2000. Tobago: Division of Finance and Planning, Office of the Chief Secretary, THA.

———. 2000a. Review of the Economy and Policy Measures, 2000. Tobago: Department of Planning, THA.

———. 2000b. Tobago Business Register 1997/1998. Tobago: Office of the Chief Secretary, THA.

———. 2001. Tobago Tourism Sector: January–March 2001. Tobago: PRDI with the Department of Tourism.

Redclift, M. 1987. *Sustainable development: Exploring the contradictions.* London: Routledge.

Revell, A., and R. Rutherfoord. 2003. UK environmental policy and the small firm: Broadening the focus. *Business Strategy and the Environment* 12 (1):26–35.

Roberts, S. 1994. *Tobago after Hurricane Flora: 1963–1973.* Unpublished BA Caribbean Study Project, University of the West Indies, St. Augustine Campus.

Roberts, S. 2000. *Community participation in tourism planning—an examination of some of the factors that contribute to participation outcomes.* Unpublished M.Sc. thesis, University of Surrey, UK.

———. 2004. *Sustainability contributions: An exploratory analysis of the small tourism enterprise in Tobago.* Unpublished doctoral dissertation, Brunel University, UK.

Rogerson, C. 2002. *Tourism, small enterprise development and empowerment in post-apartheid South Africa.* Proceedings of International Conference on Small Firms in the Tourism and Hospitality Sectors. Leeds, UK: Leeds Metropolitan University, September 12–13. .

Sasidharan, V., E. Sirakaya, and D. Kerstetter. 2002. Developing countries and tourism ecolabels. *Tourism Management* 23 (2):161–74.

Sharpley, R. 2000. Tourism and sustainable development: Exploring the theoretical divide. *Journal of Sustainable Tourism* 8 (1):1–19.

———. 2002. Sustainability: A barrier to tourism and development? In R. Sharpley and D. Telfer, eds., *Tourism and development: Concepts and issues.* Clevedon, UK: Channel View.

Shaw, G., and A. Williams. 1987. Firm formation and operating characteristics in the Cornish tourist industry—the case of Looe. *Tourism Management* 8 (4):344–48.

———. 1990. Tourism economic development and the role of entrepreneurial activity. In C. P. Cooper, ed., *Progress in tourism, recreation and hospitality management,* vol. 2. London: Belhaven.

———. 1998. Entrepreneurship, small business culture and tourism development. In E. Ioannides and K. G. Debbage, eds., *The Economic geography of the tourist industry.* London: Routledge.

Smart, B. 1985. *Michel Foucault.* Chicester, UK: Ellis Horwood Ltd.

Springett, D. 2003. Business conceptions of sustainable development: A perspective from critical theory. *Business Strategy and the Environment* 12 (2) 3:71–86.

Stabler, M. J. 1997. An overview of the sustainable tourism debate and the scope and content of the book. In *Tourism sustainability—principles to practice*. Oxon: CAB International.

Stanworth, J., and C. Gray. 1991. *Bolton 20 years on: The small firm in the 1990s*. London: Paul Chapman.

Swarbrooke, J. 1999. *Sustainable tourism management*. Oxon: CAB International.

Teszler, R. 1993. Small-scale industry's contribution to economic development. In I. S. Baud and G. A. de Bruijne, eds., *Gender, small scale industry and development policy*. London: IT Publications.

Tobago House of Assembly. 1999. Second Draft—Tobago Tourism Policy. Tobago: Tobago House of Assembly.

Tobago News. 2003, May 22. Tobago has something special.

Tosun, C. 2001. Challenges of sustainable tourism development in the developing world: The case of Turkey. *Tourism Management* 22 (13):289–303.

Tourism Development Company. June 2007. Stayover arrivals direct to Tobago. Trinidad: Trinidad and Tobago, TDC.

Van der Duim, R. 1997. The role of small entrepreneurs in the development of sustainable tourism in Costa Rica. In H. Dahles, ed., *Tourism, small entrepreneurs and sustainable development—case studies from developing countries*. Netherlands: Tilburg, ATLAS.

Vernon, J., et al. 2003. The 'Greening' of tourism micro-businesses: Outcomes of focus groups investigations in South West Cornwall. *Business Strategy and the Environment* 12 (1):46–69.

Walker, S., and H. Green. 1979. *The development of small firms policy: Planning Paper No. 14*. Leeds, UK: Planning Research Unit, Leeds Polytechnic.

Wall, G. 1997. Sustainable tourism—unsustainable development. In S. Wahab and J. Pigram, eds., *Tourism, development and growth—the challenge of sustainability*. London: Routledge.

Weaver, D. 2002. Perspectives on sustainable tourism in the South Pacific. In R. Harris et al., eds., *Sustainable tourism: A global perspective*. Oxford: Butterworth-Heinemann.

Weaver, D. B. 1996. Peripheries of the Periphery—Tourism in Tobago and Barbuda. *Annals of Tourism Research* 25 (2) 292-313.

———. 1993. Sustaining the ggo. *Journal of Sustainable Tourism* 1(2):121–29.

Willis, K. 1995. Imposed structures and contested meanings: Policies and politics of public participation. *Australian Journal of Social Issues* 30 (2):211–27.

Wood, D. 1993. Sustainable development in the Third World: Paradox or panacea? *The Indian Geographical Journal* 68, 6–20.

World Commission on Environment and Development. 1987. *Our common future*. New York: Oxford University Press.

9 'A Squatter in My Own Country!'

Spatial Manifestations of Social Exclusion in a Jamaican Tourist Resort Town[1]

Sheere Brooks

INTRODUCTION

The impact of tourism on local communities has formed a large part of tourism and interdisciplinary studies based on the cultural, social, and economic impact of the sector.[2] In the same vein, but with a more specialist focus, this chapter will explore the interaction between the tourism industry and local communities by examining the sociospatial effects of tourism development on local residents' contact with the 'tourism space.'[3] Previous studies on the spatial effects of tourism development have been explored through studies related to the tourism-built environment and have been commonly associated with the 'enclavic' model of tourism development. These discussions have centred on the destructive effects of 'enclave tourism' on local communities through case studies of countries in Africa, the Caribbean, and Latin America (Brohman 1996; Oppermann 1993; Britton 1982; Kermath and Thomas 1992; Sindiga 1996; Clancy 1999; Henshall Momson 1985).

Relatedly, this chapter also considers the relationship between the expansion of tourism and the informal employment sector. Tensions between the formal and informal employment sectors have been explored where several studies have argued that 'traditional' modes of development do not disappear because of modernization but exist alongside the capitalist economy; both sectors operate side by side (Drakakis-Smith 1987; Gilbert and Gugler 1992; Van Dierman 1997; Verschoor 1992). This duality between the two sectors has been explored in terms of the legitimacy of the informal sector operating alongside the formal sector with scant support for the informal sector exhibited by respective governments (Dahles and Bras 1999).

While this study touches on both these issues, this chapter will examine at the microlevel the interaction between the residents of a squatter settlement with the 'tourism space.' By their very existence, squatter-settlement communities are perceived to be 'illegal' and are commonly found either in or near areas of high economic activity (Baross and Van der Linden 1990; Coit 1995; Eyre 1997; Ulack 1978; Ferguson 1996). However, the frame of reference in this study is specifically concerned with the experiences of residents of a squatter settlement that has been in existence for over 40 years and is regarded as the oldest and most established of such communities close to the centre of the resort town Ocho Rios, which is the focus of this study.

BACKGROUND TO THE STUDY

The existence of deprived communities such as squatter settlements across Jamaica is one of the main consequences of the harsh economic realities that have made home ownership for the majority of Jamaicans difficult to achieve. This is compounded by the location of such communities in areas of high economic activity (UN Habitat 2003; Aldrich 1995; Coit 1995; Hardoy 1995). The existence of these communities close to tourism centres has become more precarious as Jamaica's economic dependence on tourism has increased pressure to expand and diversify the tourism product (Master Plan for Sustainable Tourism Development, Government of Jamaica 2001).

This paper therefore discusses how spatial manifestations of social exclusion have occurred in creating an exclusive 'tourism space.' It also examines how residents of a squatter settlement close to the tourism space negotiate access and interact with the tourism space in terms of income generation. Such an enquiry has implications for the 'right' of access to the designated tourism space as well as matters of social justice in terms of the 'right' to earn a living. This in turn has implications for governance, in terms of the tactics used by actors and the methods they employ in managing and controlling access to the tourism space.

SPATIAL MANIFESTATIONS OF SOCIAL
EXCLUSION AND THE 'TOURISM SPACE'

Before considering the meaning of spatial social exclusion and its application to this study, a brief overview of the social-exclusion discourse (from which the spatial aspect is derived) is undertaken. Based on a number of indicators, social exclusion occurs when a person is unable to participate in normal activities such as access to education, training, employment, housing, financial resources, and social interaction (Le Grand 1998; Lee and Murie 1999; Social Exclusion Unit, United Kingdom 1997; Atkinson, A. and J. Mills 1998; Burchardt and Mills 2000). Rather than focusing on traditional ways of measuring poverty, which tends to be restricted to a limited number of indicators over a fixed period of time, the social-exclusion framework focuses on processes that can lead to a state of prolonged poverty and ultimately a state of being socially excluded. It is this multidimensional aspect of the social-exclusion framework that has particular relevance to the case study of this chapter.

In seeking to unpack the concept's multidimensional scope, Byrne (1999) applies a 'neo-Durkeheimian' treatment to the theoretical framework by bringing to bear the question of 'identity' and arguing that social exclusion relates to 'excluded identities.' In this respect, Byrne (1999) describes 'identity exclusion' as always subject to modification by the economic relations of phases of capitalism and has an intrinsic content of its own and may exist under all forms of economic relations. Byrne (1999) argues that any form of economic activity and development will result in the exclusion of some individuals from benefits. This brings into sharp focus an interesting angle to the multidimensional scope of the social exclusion discourse, which is the interaction between economic activity along with the formation of 'excluded identities.' Byrne's (1999) contribution to the debate deserves further scrutiny here. First, the formation of excluded identities in the context of economic relations is particularly engaging because it presupposes a link between a changing economic situation alongside the shaping of an excluded identity. By this, Byrne (1999) is essentially enquiring into how excluded identities are formed during phases of economic change. It is therefore important to clarify how spatial exclusion specifically occurs in the tourism space. Spatial exclusion can occur and is a consequence of the built environment, where 'space constitutes a landscape of domination' (Sibley 1995). According to Livingstone (1992, quoted in Sibley, 1995), neoclassical economics has dictated economic behaviour, which in turn has inscribed spatial organisation. This suggests that spatial factors contribute to containing, guiding, or constraining economic and social processes. Giddens (1984:368) adds to this by arguing that the use of 'space' has to be considered in terms of its involvement in the constitution of systems of interaction. In other words, space is constituted by power relations. Sibley (1995:81) also argues that 'socio-spatial exclusion' acts as a form of social control, and social control is used in the sense here to imply an attempt to regulate the behaviour of individuals and groups by other individuals or groups in dominant positions. Applied to the context of this chapter's discussion, social control is used to sustain the tourism industry in terms of promoting actions that conform to the specificities of the tourism product.

It seems, therefore, that any study seeking to explain the spatial manifestations of social exclusion and closure should take a two-pronged approach. One such approach should consider what Sibley (1995) calls the 'normal' with the 'deviant' and the 'same' as well as the 'other.' The other approach should take into account governance tactics such as social control and power dynamics used to preserve this tourism space. It stands to reason that the governance of space relies to some extent on a measure of social control, which will have implications for the exclusion of some groups and the inclusion of other groups (Sibley 1995). This in turn bears resonance to Byrne's (1999) earlier point regarding the formation of excluded identities.

From another perspective and relevant to the governance of social space is another dimension through which spatial social exclusion can occur, this being the role of 'agency.' In this regard, agency can be thought to imply the role of individuals, clubs, and associations in a position to influence the governance and order of social space by acting as gatekeepers (Jordan 1996). Arising from this discussion, it would seem that explanations about exclusion require an account of barriers, prohibitions, and constraints on activities from the point of view of the excluded and the excluders.

The Informal Sector

Informal livelihoods are fundamental to a country such as Jamaica, which has a high level of unemployment. In the literature, there appears to be no consensus as to what constitutes the term *informal sector.* According to Mvungi and Ndanshau (2001) and Devas (2004), the most frequently adopted definition defines the sector as having the following characteristics: ease of entry, reliance on indigenous resources, family ownership of enterprise, and small scale of operation. Other descriptions are in relation to labour-intensive and adapted technology, low income, skills acquired outside the formal school system or training programmes, and unregulated and uncompetitive markets. The informality of the sector means that these are the most common channels through which livelihoods are generated. However, informal activities in the tourism sector are not usually recorded in tourism statistics.

Chambers (1995) and Lloyd-Jones and Brown (2002) define livelihoods as comprising the capabilities, assets (including both material and social resources), and activities required for making a living. This definition has relevance to a myriad of livelihood activities that can take place in public space and which is also applicable to Jamaica. Descriptions of the informal economy have been based on the notion that the sector is characterised as illegal, preindustrial, 'traditional,' and, moreover, unorganised, underdeveloped, and constituting a stagnant part of the economy (Evers 1991). In a tourist resort town such as Ocho Rios, informal activities may involve the buying and selling of goods; the selling of goods by itinerant traders on behalf of wholesale establishments; craft vendors and manufacturers; the washing of cars; informal guides; taxi-drivers; and beauty services, including hairdressing.

The informal tourism labour market plays a fundamental part in the lives of most residents living in a squatter settlement. This is particularly so in a country and in a sector of the country's economy that does not offer adequate job opportunities.[4] Because one of the pull factors of migration to an area with high economic activity is based on the promise of employment, then such a location may be the site for the activity of informal livelihoods if the formal sector is unable to absorb an excess pool of labour. The provision of employment opportunities is particularly pertinent, especially since Jamaica's Master Plan for Sustainable Tourism Development (Government of Jamaica, Master Plan for Sustainable Tourism Development 2001) has acknowledged the existence of an exclusive tourism industry that benefits the few, emphasising, as a more suitable option, an inclusive tourism industry involving the local community. If strides are to be made to widen participation in the tourism industry and for the sector to become more inclusive through the transfer of benefits to the local community, then accommodating informal livelihood activities may lead to a more equitable transfer of benefits.

Research Design

The subject of this paper forms part of a larger study that involved data collection in the town of Ocho Rios in Jamaica. The sampling and selection of respondents were based on the social exclusion theoretical framework, which recommends that explanations about exclusion require an account of barriers, prohibitions, and constraints on activities from the point of view of the excluded and the excluders (Byrne 1999). The sample of respondents in the study therefore included government representatives and actors from the tourism private sector based in Kingston and in Ocho Rios. The organisations that were sampled were the Jamaica Tourist Board, the Jamaica Hotel and Tourism Association, the Tourism Product Development Company, the Ministry of Tourism, St. Ann Development Company. and the St. Ann Chamber of Commerce.

These organisations were selected based on their role as umbrella agencies representing the interests of tourism stakeholders and entrepreneurs as well as their input in the development and planning of the tourism industry. Representatives from Non-Government Organisations acquainted with squatter settlements in St. Ann were also included in the sample. Apart from selecting response groups thought to be in a position to address the research enquiry, data from these varied sources also served to enhance the validity of the data. In addition, data collection was conducted among residents of a squatter settlement located within a two-mile distance from the centre of Ocho Rios. The methodological approach took the form of a qualitative study using a grounded theory approach for the purpose of theory generation.

The sampling and selection of respondents from the squatter settlement for interviews could not be achieved on the basis of a sampling frame. Such communities are not defined or formalised in terms of population and demographic parameters. Furthermore, the openness of squatter communities and the frequency of population movement into and out of this type of community setting made it difficult to assess the population in the absence of data based on the population. Aerial photography of the squatter settlement shown to the author confirmed the chaotic layout of the community, which in turn made any mapping exercise for the purpose of creating a sample frame difficult to contemplate.

Based on the nature of the research, it was essential to ensure that respondents were able to respond to questions posed to them. Despite

Figure 9.1   An aerial view of Ocho Rios.

the absence of a sampling frame, this researcher was keen to ensure the sample of respondents was representative of the entire population to be found from all sections of the community. This meant ensuring the resulting data reflected the views of all age-group categories as well as both gender groups.

As part of the process of familiarisation of the community, the author conducted an initial reconnaissance trip into the squatter settlement case-study area. In the absence of a sampling frame, the most useful sampling strategies that could be used were explored. The approach involved a process of observation in the community which allowed the author to familiarise herself with the general layout of the community. During this time, four entrance and exit points to the community were identified and later designated as data-collection points for conducting interviews.[5] In order to ensure the data reflected a cross section of information reflecting the diverse range of activities in the labour market, the method of 'timed sampling,' as a strategy for conducting interviews at the four main entrance and exit points of the community, was used.

So, for instance, conducting interviews during the morning captured those respondents who might be returning to the community after having taken their children to school or returning to the community. Early afternoons and evenings were deemed to be suitable times to collect data from those respondents not at work (mothers collecting their children from school or arriving back into the community from doing an errand), as well as those individuals returning to the community from formal and informal work activity.

In view of the research aims, observation data proved to be an important insight into the daily lives of the residents, which justified the use of timed sampling. Robson (1993) makes the point that the significance of conducting timed samples is based on the ability to sample across time by targeting a particular location at different times of the day or week. Timed sampling was complemented by a system of screening to verify the residency and period of residence in the community. As this was an 'open' community without any precise record of residents, and neither assigned house nor lot numbers, then this was considered to be the only way in which to check whether a potential interviewee actually resided in the community.[6] Steps were taken to ensure there was a balance in the sample in respect to age-group categories and gender. Hence, checks were made to verify a respondent's inclusion in the study and, in short, ensuring the sample reflected a cross section of the population in the community as far as was possible. It was felt that the interaction of these attributes would generate a richness of data comprising varying degrees of interaction with the tourism space. The resulting sample of respondents interviewed for the study consisted of fifteen female and fourteen male respondents.

Because the aim of the study was to gain an insight into differences of experience among respondents in the formal and informal labour market in terms of their contact with the tourism space, attributes such as age and gender were crucial to disaggregating the resulting data according to these attributes. In aiming to ensure the resulting data were derived from a cross section of the community, interviews were conducted iteratively so that once theoretical saturation had been achieved in one section of the community, data collection would commence in another section of the community. Ultimately, the use of the timed sampling approach was important to increasing the analytical potential of the data.

THE STUDY AREA: OCHO RIOS

This section will outline the basis on which the resort town of Ocho Rios was selected as the study area. The tourist resort towns of Montego Bay, Negril, and Ocho Rios are the three main tourist destinations in Jamaica; they offer the mass tourism product and feature coastal-based resorts and hotels with high capacity accommodation facilities (Jayawardena 2002). Negril is the smallest resort town and differs in some respect from the other two towns in terms of offering a combination of mass tourism (on a limited scale, although tourism statistics indicate increased development in this sector) with community tourism. The resort town is characterized by small villas and guesthouses operated by local people, and this is the only resort town where tourism was initiated by local people, as opposed to the large-scale tourism development by external agencies (McKay 1987).[7]

The resort town of Ocho Rios offers a combination of mass tourism and some nature-based tourism attractions and small guesthouses to be found in the hillsides overlooking the town. The town also has a buoyant cruise-shipping industry. According to travel statistics (Ministry of Tourism 1997–2004), Ocho Rios has continued to provide the larger share of Jamaica's cruise arrivals. The Ocho Rios port accounted for 810,000, or 72 percent, of the 1,132,596 passengers arriving in Jamaica, with Montego Bay increasing its share from 20 percent in 2002 to 28 percent, with 319,405, in 2003. Similarly, hotel occupancy levels by area showed that in both 2002 and 2003, Ocho Rios surpassed Montego Bay as having the highest level of hotel room occupancy. According to the Master Plan for Sustainable Tourism Development (2001), Ocho Rios attained the position of dominance formerly held by Montego Bay in regard to employment levels in the tourism industry. The master plan also notes that the rate of growth in tourism accommodation in the Ocho Rios area is increasing at a much faster rate than that occurring in Negril, with the number of rooms increasing from 2,733 in 1982 to 5,782 in 1992. The Ocho Rios area had 29.5 percent of the island's room capacity in 1996, up from 26 percent in 1982 (Master Plan of Sustainable Tourism Development 2001).

Population census data indicate that as a town with tourism activity, Ocho Rios was categorized as one of the fastest-growing urban towns. It is noticeable that as the island's major tourist centre, Montego Bay is not in this category. In a comparison of population movement in all of the main tourism destinations, Ocho Rios had the highest percentage rate of population growth between the census periods 1991 and 2001 (Population Census, Jamaica Country Report 2001).

The existence of squatter settlement communities

The existence of squatter settlements in tourist resort towns has been cited as a pervasive problem (Eyre 1997; Ferguson 1996). The extent of the problem in terms of the prevalence of such communities in close proximity to the centre of Ocho Rios was never fully assessed through statistical records because no comprehensive database currently exists. This information was therefore gleaned through formal and informal contact with officials at the Ministry of Water and Housing, the local parish council, the St. Ann Development Company, and Relocation 2000 (a subsidiary of the National Housing Trust, which is the government's housing-development agency). The government's adviser on squatting from the Ministry of Water and Housing only surmised that 'tourism stakeholders in Ocho Rios were scared at the prospect of Ocho Rios having as severe a problem of squatting as that taking place in Montego Bay' (Interview with respondent in February 2005).

The extent of squatter settlements was assessed using official statistics from UN Habitat; the Jamaican Survey of Living Conditions (JSLC) and various years of Jamaica Population Census—it would not have been appropriate to use JSLC alone as this is not representative at the parish level in Jamaica (UN Habitat) .[8] Boxill (2002) indicates that as the tourism economy expands, the prospect of jobs in the industry encourages migration into the resort town. Montego Bay and Ocho Rios have larger squatter settlements, with some existing before the start of tourism development in both towns. Tourism development in Ocho Rios features a high density of tourism and commercial businesses clustered together in the centre of the town (the plains), with squatter settlements situated in the hilly areas overlooking the town centre (Master Plan for Sustainable Tourism Development 2001). These hilly areas also happen to be the location of several nature-based tourism attractions.

Criteria used in the selection of the Parry Town squatter settlement

Research Methods

Open-ended interviews using a semistructured schedule were used for the collection of data from all response groups in the sample. In addition to this, the participant-as-observer method was used as a method for eliciting data through the author's attendance at resort-board meetings, held under the auspices of the Tourism Product Development Company (TPDCO). The significance of assuming the role of *observer-as-participant* was instrumental in collecting data that could be matched with data derived from other sources. This complied with Jorgensen's (1989) point that 'observation' is appropriate when there are important differences between the views of insiders as opposed to outsiders. Observations of different settings required what Jorgensen (1989) describes as a logic and process of inquiry that is open-ended, flexible, opportunistic, and requires constant redefinition of what is problematic based on facts gathered in concrete settings of human existence.

Administrative data from tourism statistics, literature sources, and government policy studies, along with photographs and Internet sources of information, were also consulted. The various methods of data collection were used as a strategy for accumulating rich data, both for the confirmation of results and for planning the continued process of data collection. This resulted in a triangulation of data emerging from various sources, which, in turn, systematically increased the reliability and validity of findings.

The model of tourism development and the creation of spatial exclusion in Ocho Rios

The first strand of the research enquiry sought to assess, through interviews with key informants, the underlying factors that contribute to the development of sociospatial exclusion. . This strand of the enquiry produced data that could be used to assess the impact of the planning and design of the resort town in creating conditions relative to sociospatial exclusion. The findings showed that an enclavic model of tourism development had occurred, and this continues to be the case (Oppermann 1993; Kreth 1985; Henshall Momsen 1985; Jayawardena 2002; Patullo 1996).

Although there are plans for diversifying the tourism product and taking tourism entities into areas beyond an already clustered town centre, this has invariably resulted in a clustering of tourism attractions and accommodation establishments situated in the vicinity of the coastal zone and in the centre of the town of Ocho Rios.

Views were expressed suggesting that the process of development was not well thought of, thereby implying a mismanagement of the planning process:

> They really didn't have to put everything along the same strip (facing the sea) although I suppose the beach front was the main attraction and advantage had to be taken of that. . . . I think that Ocho Rios is in a real mess! (Respondent—Non-Government Organisation, Ocho Rios)
>
> I don't think that there was an original intention to transform this town into a high rise tourist area as it was never allowed to grow slowly out of its sleepy fishing village image. (Project Manager, Tourism Product Development Company [TPDCO])

As can be seen, the priority of encouraging tourism investments in the resort town more than likely surpassed a structured system of town planning. This paints a picture indicating that the spatial organisation of the resort town is closely aligned to the geographical lanscape, suited to tourism. This development is also driven by private investment keen to ensure competitive advantage through the spatial location of businesses in this vicinity. Such a situation raises questions about the role of local government authorities in governing this space and in the planning and design of this resort town. The literature suggests that there is likely to be tensions between central and local governments, especially as tourism tends to be the preserve of central government (Duncan 1990; Jones 1996). Hence, the function and role of the local government machinery is either regarded as nonfunctioning or, at best, is completely disregarded by dominant actors in the tourism sector.

The weakness of the local government machinery has created the space in which the tourism private sector has a controlling interest in decisions pertaining to the planning of Ocho Rios. Evidence did arise during the course of this study, when a respondent from a local environmental NGO, who is also a member of the civic group, St. Ann Parish Development Committee

(PDC), remarked, 'We were asked as a committee to comment on the environmental impact assessment for the hotel that is going up on the Mammee Bay lands and was told to send in our comments within a week, but this request was sent to us eight weeks after construction had started, so why ask us to make a comment, when construction had already started!' The influence of gatekeepers from the tourism private sector was evident when they expressed their concerns about the ongoing pace and direction the current programme of tourism development was taking:

> 'Nobody has really planned beyond the first set of work that the Urban Development Corporation did towards developing the town into a tourist resort town. Now it needs that second phase of development to consolidate what has already been done' (Managing Director, Coyaba River Gardens and Museum).

As investment in tourism development has been driven largely by the private sector, with government in the role as facilitator, this may have created a vacuum through which tourism gatekeepers have been able to influence the spatial structure of the resort town. Hence, the large number of businesses and tourism entities clustered in the centre of Ocho Rios makes the management of the tourism space easier to monitor and control. A problem with this clustering as a contributor to sociospatial exclusion was raised in connection with the existence of all-inclusive hotels. It was thought that the clustering of these entities served to create a boundary between these hotels and local communities. It became apparent that some stakeholders were of the view that the all-inclusive hotel chains have had their day. The task at hand was to take the tourism product to the outlying areas of Ocho Rios, where the scope for diversifying the tourism product could be achieved and which could afford the visitor an alternative and more diversified visitor experience:

> Tourism has now gotten to a point where if we didn't have the all-inclusives, then we couldn't have tourism because they provided the security that the guests craved for. It's time now to take another look at it and I think we ought to curtail all-inclusive hotels and move more into community and environmental tourism. (President, St. Ann Chamber of Commerce)

The views of respondents seemed to suggest that diversifying the tourism product could stem the number of all-inclusive hotels because visitors might be more tempted to embark on a more diverse holiday experience in Jamaica: 'When I visit a new country I want to see how the people live . . . the all-inclusive hotels are so well packaged that the people are almost prisoners of their own vacation! I would want to see more of the country' (Mayor of St. Ann Parish Council).

The existence of the all-inclusive chain of hotels that concentrates visitors to the island in a confined space may be limiting the opportunity for diverse holiday experience. The literature on the interaction between the all-inclusive hotel establishments and the local community has been debated at length in regard to the trickling down of economic benefits into the local community (Patullo 1996; Dunn and Dunn 2002; Jayawardena 2002; Kreth 1985; Boxill and Maerk 2002).

Restricting the tourist to the confines of the hotel complex does in itself create a spatial barrier between the tourist and the local community; at the same time, it creates a perceived demarcation between the tourism and no-tourism sections of the town. The expansion of tourism beyond the centre of Ocho Rios was seen by three respondents as vital to reducing the importance of the all-inclusive hotel establishments and their impact in the centre of the resort town: 'The time has come for us to expand outside of the town . . . it is just too saturated here now. We have too high a concentration of all-inclusive hotels here!' (Director, St. Ann Development Company).

Taking tourism outside of the centre of Ocho Rios included the possibility of moving into nearby towns such as St. Ann's Bay, where the prospect of offering a diversified tourism product in the form of heritage tourism could be achieved: 'I would like to see the development of St. Ann's Bay as a tourism site, but we are just talking and nothing else is happening' (Chief Executive Officer, St. Ann Parish Council).

The prospect of taking the tourism product outside the centre of Ocho Rios to the hilly terrain overlooking the centre of Ocho Rios and to other small towns in St. Ann was considered as one way of diversifying the tourism product while offering more attractions

TOURISM DEVELOPMENT AND THE INFORMAL ECONOMY

The aerial view of the town of Ocho Rios shows a dense grouping of tourism entities in the centre of the resort town, with a clustering of squatter communities located on hillsides overlooking the centre of the town. Moreover, scant regard has been made to integrate informal trading activity, such as indigenous craft markets, with formal businesses. Where this does occur, as illustrated in Figures 9.1, 9.2, and 9.3, the craft market directly faces an in-bond shopping mall, with the former having an unattractive appearance in contrast to the latter. Second, there are comparatively few such outlets for informal traders located in the main tourism centre.

In the area of main tourism activity and with the heaviest tourism traffic in respect to visitors, there are only two official craft markets. When compared to the comparatively large number of in-bond shopping malls to be found in this vicinity, these informal markets are few. There are two smaller markets located in the outskirts of this area of main tourism activity, but these are not exposed to the tourism traffic to be found in the centre of the town. In addition to this, the concentration of similar businesses intensifies the level of exclusion from the city centre of itinerant and informal traders from the more lucrative trade to be obtained from large tourist groups.

Figure 9.2   An open-air craft market in Ocho Rios.

Figure 9.3   The entry to a craft market in the town.

The visual appearance of a tourism shopping outlet is a key indicator to attracting business. The attractive appearance of an in-bond mall, in comparison to a dilapidated craft market, provides competitive advantage for the in-bond mall in respect to sales and visitor numbers venturing into such a setting. It was observed that in the craft market, there were no formal display systems for showcasing goods, and so vendors resorted to displaying their wares on plastic sheeting laid out on the ground (see Figures 9.2 and 9.3).

On the other hand, the immaculate appearance, barriers and gates, and the imposing presence of security guards serve to reinforce the exclusivity of the in-bond shop malls as well as setting boundaries around these establishments (see Figure 9.4). The visual appearance can be construed to be an avenue through which exclusion from livelihoods could be achieved, thereby developing into financial exclusion from potential markets on the part of the craft vendors and manufacturers.[9] (see Figure 9.5)

As plans continue to develop the town centre's appeal, a respondent from the Tourism Product Development Company (TPDCO) spoke about plans to develop a heritage park in the near future, where it was originally thought that formal and informal traders could be housed at one location to showcase Jamaican-made products, but this plan has since been discarded. The comments of this respondent are noteworthy:

Figure 9.4 A shopping center with a sign that reads "Duty Free Shopping Soni's Plaza Downtown Ocho Rios."

Figure 9.5 A woman standing by a sign that reads "Soliciting by braiders, gigolos, taxi drivers, etc., strictly prohibited on this property by order of management."

We have plans for a park called the 'One Love Trail,' which will be a heritage park and a walking attraction depicting Jamaican art. We're looking at small specialist boutique shops featuring vegetarian shops, natural jewellery etc. One would want to stop at these shops as no two shops are the same, but not what you would find in the craft market at all! (Project Manager, Tourism Product Development Company)

Hence, there appears to be a 'locking out' of informal trade not just by formal businesses but also by government agencies charged with the remit of planning and developing the tourism product. A derivative from this clustering of prime businesses is the ability of traders and the tourism private sector to collectively protect and sustain their assets through mechanisms such as civic groups and linkages with the local parish council and other agencies involved in decision making. This would invariably place them in a position to act as gatekeepers in scrutinising the admittance of other traders into this vicinity. Using social capital and networking strategies, such groups are in a position to influence decision-making processes in regard to the design and planning of the resort town.

The town has a number of influential business-support and civic groups (comprised of tourism and nontourism actors) largely made up of business stakeholders in the tourism industry. So, for example, the Jamaica Chamber of Commerce, the Jamaica Hotel and Tourism Association, the Indian Merchants Association, and the Ocho Rios Civic Group, with the involvement of members from the St. Ann Parish Development Committee, all play a fundamental role in governing the 'tourism space.' Other small civic groups exist and, according to the comments of one respondent, 'Hardly a week goes by when there isn't another civic group that comes into existence here in Ocho Rios.' [10] The influence of these groups is fundamental to maintaining the status quo by ensuring that only those persons or businesses in line with this status quo are able to enter this 'club.' Without the financial clout or required assets, the exclusiveness of the tourism space would be unable to accommodate individuals without this requirement. Because the control of the business area in the main tourism space of Ocho Rios is governed by few organisations, this presents a challenge for informal traders who lack the required assets to enter and function as entrepreneurs in this zone.

In view of the state's severe limitation as investors in the tourism industry, this gives elites from the tourism private sector the forum through which to influence decision-making processes relating to the governance of space in the tourist resort town. Part of the reason for this vacuum stems from the weakness of local government institutions such as the local parish council that has not instituted planning controls. It has been argued that the devolving and decentralisation of power from central to local government is critical for effective spatial tourism planning. However, the implications of what the devolution of power means in terms of the type of participants and their level of influence in the decision-making processes have made this a difficult policy objective to realise (Duncan 1990; Jones 1997). Effort has been made to institute a 'bottom-up' approach through the development of the Parish Development Committees (PDC) across all the parishes, with the mandate of acting as a local governance mechanism with links to the local parish council.[11] Still, there is no indication that these local organisations have taken sufficient account of, or made allowances for, those operating in the informal sector.

While attending two such meetings of the St. Ann Parish Development Committee, the author observed that the first meeting mainly had the attendance of community representatives from the parish as well as representatives from two environmental NGOs. While the second meeting consisted of the same complement of attendees, a proprietor of a nature-based tourism attraction was there seeking support for his efforts to deal with an environmental problem in the area of his business. In the case of the last meeting, the tourism entrepreneur invariably dominated proceedings in this meeting by highlighting his concerns and he left the meeting before it had finished. Hence, such meetings can be controlled by players in the tourism industry who are in a position to manipulate proceedings while seeking to elicit support from others linked to the tourism industry and from local communities. As this is a key civic interest group based in the parish of St. Ann, where the main tourist resort of Ocho Rios is located, it is likely that tourism interests would be able to manipulate and gain legitimacy for the protection and expansion of the tourism sector.

The perception that barriers existed to prevent entry into the centre of Ocho Rios was felt by residents living in the squatter settlement community to be of concern, especially, for the most part, those dependent on making a living in this tourism setting. Attempting to break this barrier by merely entering the 'tourism space' presented challenges for respondents. Two respondents made the following comments:

> I try not to talk to the tourists as if you talk to them the police are ready to arrest you for harassing them, but I have seen some of the tourists come up here and talk to us. . . . I don't deal with tourists at all as I can't understand them and I cannot stand the Police as I don't like trouble. (Male respondent, 21 years of age, Parry Town Squatter Settlement)

> I am walking in Ocho Rios and a tourist stops me and asks for information and a police will see me . . . and tell me to leave the lady alone, why am I pimping her and they are ready to lock me up for abuse, and for harassment of tourists. (Female respondent, 40 years of age), Parry Town Squatter Settlement)

These findings are consistent with those that have emerged from a study done by Dunn and Dunn (2002) in which respondents expressed a sense of prohibition from entering the main tourism town for fear of being harassed by the authorities, if found to be in contact with a tourist. In connection with this, the proprietor of a tourism attraction in the town made the following comment: 'I employ some people who live in the squatter settlement and believe me, they cannot come down here to Ocho Rios for their pay because they are afraid of the police stopping them. They have no form of identification on them' (Interview with the owner of a tourism attraction in Ocho Rios, February 2005).

On the other hand, it was interesting to note that exclusion from the resort town was not restricted to social interaction in the resort town's centre, as identified by a formal tourism worker living in the squatter settlement:

> I have had guests from Germany and Denmark coming here (to the squatter community in Parry Town) as they were interested in seeing where I live. I would take them around this place and they would be struck by the various trees and insects here. When tourists come here some don't want to stay in a hotel but wish to move around and meet the Jamaican people and know what its like to live in Jamaica.[12] (Female Community Leader, squatter settlement in Parry Town)

The respondent in this case took pride in showcasing her community (despite it being a squatter settlement) and her country. The fact that this respondent lives in a squatter settlement was of no consequence to her and could be attributed to the fact that she did not see her living environs as a 'prohibited area,' but this was in fact her home and obviously a source of pride. Tourism stakeholders involved in the planning and management of the resort view the likely entry of tourists into squatter areas as justification for pursuing an 'enclavic' model of tourism development. Such planning strategies are based on setting up barriers in the tourism space that do not make provision for the aspirations of some of these excluded groups to participate in the tourist industry

The prevailing view of some tourism stakeholders is based on the notion that the tourism industry has to be protected, and the tourist should be shielded away from the locals and from undesirable locations. Hence, for the tourist choosing to venture outside of this periphery, notions of prohibition may not be enforced and do not represent forced exclusion. However, this is not to say that the tourist might not have been warned away indirectly from venturing into areas considered to be 'no-go.' Prior to even arriving in the island, would-be visitors may have been made aware of travel advisories issued in their country through travel agencies concerning these sites.

However, for locals wishing to venture into the exclusive zone of the tourism space, there are restrictions. In the resort town of Ocho Rios, tourism space has become 'sacred,' which prohibits locals from daring to venture into its environs. Sibley (1995) expresses the view that boundaries between the consuming and nonconsuming public are strengthening, with nonconsumption being constructed as a form of 'deviance' while spaces of consumption eliminate public spaces in city centres. A large part of the population in Jamaica is heavily dependent on livelihoods from informal activities. The implications for making a livelihood in the tourism space raise the stakes even further for a large proportion of the population heavily dependent on making a livelihood informally. One respondent spoke of the difficulties encountered when conducting her business of 'buying and selling' in the centre of the town, when she said:

> I can't get any jobs at all, so I do the buying and selling which doesn't go too well as the police are always around the town and if they catch you selling, they can arrest you and seize your goods. I would like to go to the craft market and sell, but it's full . . . the police get rough if they see you at the market selling. (Female respondent, aged 33, Parry Town squatter settlement)

The importance of the informal sector for making a livelihood was seen in the number of respondents from the squatter settlement who declared themselves as not working on a formal basis in the tourism sector. Many declared they were engaged in 'hustling' activities (in the case of the male respondents) and 'buying and selling' activities (in the case of female respondents). In the sample, which was almost split evenly between male and female respondents (N = 40), over 90 percent of the sample declared this to be the case. This raises questions as to whether the exclusivity of the tourism space bars such activities. Furthermore, this brings into focus exclusionary measures used to preserve this space by actors, thereby denying opportunities for generating informal livelihoods from a sector vital to a large part of the town's population.

DISCUSSION AND CONCLUSION

The findings have established the occurrence of sociospatial exclusion through the impact of the built environment. More pertinent to this study, the findings have shown that the governance and social control of the tourism space are dominated by actors keen to control activities while setting boundaries in order to demarcate the tourism space from other localities. It would appear therefore that informal traders have become 'excluded identities' in terms of being barred from entering and, by extension, attempting to generate a livelihood in this setting. The exclusivity of the tourism space in this regard bars informal traders without the necessary assets to be admitted and to function in this locality.'[13] Several studies speak about the dilemma the informal sector in developing countries face in functioning in public space. Lloyd-Jones and Brown (2002) have pointed to the difficulty of raising capital, the single most pressing problem facing informal sector businesses. From another perspective, Nnkya (2003) speaks about the problem as having to do with the design of cities, which are conceptualised as places of formal economies. The strategy of urban design that characterises industrialised countries has been transferred to developing countries. The aforementioned authors argue that the legislation and policies in place largely provide a framework for managing cities built around such formal economies.

The challenges experienced by the informal sector could also be attributed to perceptions about the informal sector, which is viewed as a temporary measure and not necessarily a sustainable means of generating incomes. The findings in this study have shown that this can be attributed to a number of factors, one of these being the growing influence of the tourism private sector in decision-making processes, from the local government level and through to local interest and civic groups. This is in turn strengthened by the existence of a vacuum left by the state's gradual withdrawal and limited role as 'facilitator' in tourism planning in the resort town. The surge in the number of local interest groups with representation from tourism-affiliated entities and the transference of their interests through these interest groups will influence the planning of the town's tourism plant and infrastructure while monitoring and protecting investments in the tourism space. This finding is in keeping with Jordan's (1996) observation regarding the influence of the 'club' structure in creating exclusion and, once more, financial exclusion for those entities without the assets to join this 'club.'

Second, the findings show that sociospatial exclusion has the potential to create exclusion according to 'gender.' In this regard, Byrne's (1999) earlier edict, that the creation of 'excluded identities' is a derivative of spatial transformation, can have differential impacts where gender is concerned. While respondents from the squatter settlement expressed a sense of prohibition from entering the tourism space, the findings show that restriction is even more severe in the case of male respondents. Female respondents may have been aware of possible difficulties arising from entering the tourism space, but this has not necessarily stopped them from entering the town. Male respondents were more likely to not venture into the centre of the tourist town because the repercussions for them could have been even more detrimental. Hence, the marked differences in experiences in regard to gender would suggest that exclusion from employment and livelihoods for young men will have implications for their presence in the resort town with the possibility of their turning to unconventional means of making a livelihood.

The study shows that there continues to be a gulf between the formal and informal sector, which, in the light of Jamaica's Master Plan for Sustainable Tourism Development (2001), may prove to be a constraint in meeting the objective of encouraging an inclusive tourism industry. The same policy document advocates the diversification of the tourism industry, and, again, this is premised on the grounds of including the local community to achieve this goal. Yet this raises the question of precisely where the informal sector in terms of traders and craft people are accounted for in this scheme.

These strides may necessitate an 'unpacking' of the term *inclusive tourism industry* as illustrated in the policy document, if an understanding of the role of the informal sector is to be achieved as part of the objective. Once more, the policy document lists, among its objectives, the diversification of the tourism product, which suggests the inclusion of other forms of holiday experiences, which includes nature-based tourism. While this might be ideal in terms of taking tourism into local communities, this plan could have opposite implications in terms of effectively keeping locals (apart from formal tourism workers) from entering the official tourism space. In the meantime, tourists would have the option of exploring further beyond their hotel environs (this would more than likely come under the scrutiny of the hotels and tour operators).

If the notion of a designated tourism space is to be obliterated and put in its place a more 'inclusive' tourism industry, then recognition of the informal sector made up of small traders and local craft vendors needs to be incorporated as part of this endeavour. As a country heavily reliant on earnings from the tourism sector and renowned for its people being pleasant and friendly, and having a natural disposition towards tourism, the inclusion and interaction between locals and tourists need to be considered. As mentioned earlier, this is a matter of social justice, and especially in a country where the informal sector is significant in generating livelihoods for a large section of the population. Tourism promotion agencies such as the Tourism

Product Development Company (TPDCO) needs to employ strategies for alerting informal traders about a rapidly changing travel market and how best to cater to this market.

As the main driver of policy and planning, the state needs to work towards empowering these traders in seeking to set up cooperatives and small and medium enterprises (SME) in which the necessary tools for creating products while updating skills can be furnished. Institutions could be established for assisting these cooperatives with small-business loans, with flexible loan policies that are mindful of the peculiarities of the informal traders.

Second, there appears to be room for niche marketing in respect to the souvenir market. Many souvenir products are not manufactured in Jamaica (for the most part, these products are sold in the shopping malls) but are imported from countries in the Far East.[14] Most of the materials used to manufacture these goods can be found in Jamaica. Small and medium enterprises and, particularly, agribusinesses need to be fostered in local communities located in the vicinity of the resorts, and especially those communities situated on the border of resort towns. If the diversification of the tourism product is to be realised, then there could be good opportunities for linking tourism with other sectors in the economy such as organic agricultural farming industry. While the export of traditional staple agricultural products such as sugar and bananas from Jamaica has been in decline over the last twenty years, a renewed thrust could be made to create linkages between tourism and the agriculture sector. Furthermore, the travel market should be the instigator through which to formalise and create legitimacy for the informal sector. This could, in many ways, transform the concept of the 'informal labour market' while removing the stigma and necessity to do hustling, an activity mostly dominated by young men.

Socio-spatial exclusion, which characterises the 'tourism space' and the creation of 'excluded identities,' is a direct result of the mass tourism product that dominates Jamaica's tourism industry. As it is, the existing tourism space in Ocho Rios is constrained because of the scarcity of space. Actors in the tourism industry in Ocho Rios advocate a second phase of tourism development for the town that involves embarking on community tourism outside the immediate outskirts of the town's centre. Many residents in these vicinities are engaged in informal activities as a strategy for making a livelihood, so policies to integrate the informal sector will need to be contemplated. The term *formalisation* has to be qualified here. Dahles (1999) is of the view that the informal sector will survive and find ways to continue surviving. However, informal activities in the case of Jamaica involve hustling, which is dominated by young men. Jamaica's human development capacity cannot be premised on hustling activities and, more so, as the island has a young population. 'Informal' can be linked to 'excluded identities,' which in turn implies the exclusion from livelihoods. The findings in this study will have implications for other developing tourism areas in Jamaica such as Port Antonio and the South Coast region of St. Elizabeth.

In the case of Port Antonio, tourism is being resuscitated, but with plans for a low-impact, nature-based tourism being contemplated. The South Coast and St. Elizabeth are also moving in the same direction, but with community tourism being emphasised. Although the tourism products offered in these regions are different from those offered in Ocho Rios, the matter of spatial planning in regard to tourism will need to inherently reject the idea that tourism requires a consigned and exclusive 'space' for its existence. Social interaction and the generation of livelihoods in any space characterised by economic activity must be cognisant of the livelihood potential of the local community while altogether removing nuances suggesting that tourism activity requires, for its survival, demarcation and exclusivity.

NOTES

1. This was a comment made by a resident from the Parry Town squatter settlement in Ocho Rios, who challenged the notion of being referred to as a 'squatter,' even though the researcher was very careful to not use the term. In her opinion, citizenship of Jamaica could not be implicated as being a squatter.
2. In this regard, references are made to studies by De Kadt 1979; Dunn and Dunn 2002; Henshall Momsen 1985; Jayawardena 2002; and Patullo 1996.
3. Some of these reasons were based on government agencies being unable to identify alternative locations for relocating these communities. Some commentators cited 'political clientelism' as a possible reason for the community not being relocated.
4. It is difficult to measure overall labour-market participation in the tourism sector because informal activities in the tourism labour market have been difficult to measure, but they form a fundamental part of the tourism labour market. The tourism sector tends to be subsumed under the services sector in official statistics. Annual travel statistics show that employment in the accommodation sector has increased only marginally across all the main tourism localities across the island (Annual Travel Statistics 2004, 2005; Jamaica Tourist Board, Jamaica). Additional statistics suggest that, to date, of the 1.2 million members of the Jamaican workforce, 47 percent of them are based in the service sector (Economic Analysis of Tourism in Jamaica 2005; government of Jamaica and the Organisation of American States. http://www.oas.org).
5. Pilot tests revealed that respondents were more willing to be interviewed when returning to the community because they had more time in which to do this, as opposed to when they were leaving the community.
6. Residence was confirmed when a resident would invite the author to his or her house where the interview could take place. The author would only opt for doing so if it was a female respondent.
7. A similar venture has been undertaken more recently in the parishes of St. Elizabeth and Trelawny, where local communities initiated what is called 'South Coast' tourism during the early 1990s. This was initiated by local communities and started off as a 'Jamaican homecoming' event to attract Jamaicans overseas (including second- and third-generation Jamaicans) to visit the island. The event coincided with a food festival featuring Jamaican cuisine.
8. The Jamaica Survey of Living Conditions (JSLC) is a nationwide survey to establish measures of household welfare and to monitor the impact of Jamaica's Human Resources Development Programme on health, education, and nutrition. When first conducted, the survey formed part of the World Bank's

Living Standards Measurement Study (LSMS) and was conducted on an annual basis. The purpose of these surveys is to provide household-level data for evaluating the effects of a variety of government policies on the living conditions of the population.

9. This is compounded by the very nature of squatter settlements, which is characterised by individuals taking control of land and, within a very short space of time, creating a settlement. Equally, a squatter settlement can be removed within a short space of time resulting from state intervention.

10. It is interesting to note that the Jamaica Tourist Board Web site fails to indicate the precise location of craft markets in the town of Ocho Rios, in comparison to the descriptions given to formal shopping malls.

11. Comments by a business proprietor from the Ocho Rios Shopping Centre in an interview held in February 2005.

12. In reference to this, the researcher had the occasion to attend and observe meetings of the local branch of this committee and found that, among local politicians and councillors, NGOs, and the proprietor of a tourism attraction (who, incidentally, attended to air his concerns about a squatter settlement situated next door to his business), attendance revealed that only local residents from the CBOs of affluent and 'legal' residential communities scattered around Ocho Rios were there. What appears to be informal links between local councillors and lower sections of the population, the concerns of the local population are not filtered to such meetings. This is, of course, a component of the larger research project and is only mentioned in the context of this paper as a point of interest and reference.

13. In some cases, this same respondent reported seeing tourists 'accidentally' walking into her community, and she and other relatives would give a guided tour of their community while addressing enquiries about the vegetation. It is uncertain whether these visitors were referred to the community with the view to obtaining marijuana or other forms of hard drugs!

14. Of course, exclusion from jobs in the formal tourism industry could be based on poor levels of education, the address of a resident (this was not the remit of this research study), ill health, and a lack of working experience, among other possible factors.

15. In this regard, it is not unusual to find plantain chips and chips made from exotic fruits imported from countries such as Thailand and Malaysia, which, similar to Jamaica, are in tropical regions and grow the same products.

REFERENCES

Aldrich, B. 1995. *Housing the urban poor: Policy and practice in developing countries.* London: Zed Books.

Atkinson, A.B., and J. Hills. 1998. *Exclusion, employment and opportunity.* London: LSE.

Baross, P. , and J. Van der Linden. 1990. *The transformation of land supply systems in Third World cities.* Aldershot, UK: Avebury.

Boxill, I., and J. Maerk. 2002. *Tourism and change in the Caribbean and Latin America.* Kingston: Arawak Publications.

Bringing Britain together: A national strategy for neighbourhood renewal. 1998. London: Social Exclusion Unit: Cabinet Office.

Britton, R. 1982. The political economy of tourism in the Third World. *Annals of Tourism Research* 9:331–58.

Brohman, J. 1996. New directions in tourism for Third World development. *Annals of Tourism Research* 23 (1):48–70.

Burchardt, T., and J. Hills. 2000. Unpublished paper to the Scottish Executive on social exclusion. London: London School of Economics and Political Science.

Byrne, D. 1999. *Social exclusion*. Buckingham, UK: Open University Press.

Chambers, R. 1995. Poverty and livelihoods: Whose reality counts? *Environment and Urbanization* 7 (1):177–204.

Clancy, M. 1999. Tourism and development: Evidence from Mexico. *Annals of Tourism Research* 26 (1):1–20.

Coit, K. 1995. Politics and housing strategies in the Anglophone Caribbean. In *Housing the urban poor: Policy and practice in developing countries*. B. Aldrich and R. Sandhu, eds. London: Zed Books.

Dahles, H. 1999. Tourism and small entrepreneurs in developing countries: A theoretical perspective. In *Tourism and small entrepreneurs: Development, national policy and entrepreneurial culture. Indonesian cases*. H. Dahles and K. Bras, eds. New York: Cognizant Communication Corporation.

Dahles, H and Bras (eds) 1999. Tourism and small entrepreneurs: Development, national policy and entrepreneurial culture. Indonesian cases: New York: Cognizant Communication Corporation.

Devas, N. 2004. *Urban governance, voice and poverty in the developing world*. London: Earthscan Publications.

Drakakis-Smith, D. 1987. *The Third World city*. London and New York: Routledge.

Duncan, N. T. 1990. The political process and attitudes and opinions in a Jamaican parish council. In *A reader in public policy and administration*. G. Mills, ed. Kingston: Institute of Social and Economic Research, University of the West Indies.

Dunn, H,, and L. Dunn. 2002. *People and tourism: Issues and attitudes in the Jamaican hospitality industry*. Kingston: Arawak Publications.

Evers, H. D., and O. Mehmet. 1995. The management of risk: Informal trade in Indonesia. *World Development* 22 (1):1–19.

Eyre, A. 1997. Self-help housing in Jamaica. In *In self-help housing, the poor and the state in the Caribbean*. R. Potter and D. Conway, eds. Jamaica/ Barbados/Trinidad and Tobago: The Press University of the West Indies.

Ferguson, B. 1996. The environmental impacts and public costs of unguided informal settlement: The case of Montego Bay. *Environment and Urbanization* 8:171–93.

Giddens, A. 1984. *The constitution of society*. Cambridge, UK: Polity Press.

Gilbert, A., and J. Gugler. 1992. *Cities, poverty and development: Urbanisation in the Third World*. Oxford: Oxford University Press.

Hardoy, J. 1995. *Squatter citizen*. London: Earthscan.

Henshall Momsen, J. 1985. Tourism and development in the Caribbean. In *The impact of tourism on regional development and cultural change*. E. Gormsen, ed. Mainz, Germany: Geographisches Institut Johannes Gutenberg-Universitat.

Jamaica annual travel statistics. Kingston: Jamaica Tourist Board: Corporate Planning and Research Department.

Jamaica master plan of sustainable tourism development. 2001. London: Commonwealth Institute.

Jayawardena, C. 2002. Mastering Caribbean tourism. *International Journal of Contemporary Hospitality Management* 14 (2):88–93.

Jones, E. 1996. Symposium on local government reform: Democratization from below. Kingston: University of the West Indies, Department of Government.

Jordan, B. 1996. *A theory of poverty and social exclusion*. Oxford: Polity Press.

Jorgensen, D. 1989. *Participant observation: A methodology for human studies*. London: Sage.

Kermath, B. M., and R. N. Thomas. 1992. Spatial dynamics of resorts, Sosua, Dominican Republic. *Annals of Tourism Research* 19:173–90.

Kreth, R. 1985. Some problems arising from the tourism boom in Acapulco and the difficulties in solving them. In *The impact of tourism on regional development and cultural change*. E. Gormsen, ed. Mainz, Germany: Geographisches Institut Johannes Gutenberg-Universitat.

Lee, P., and A. Murie. 1995. *Literature review of social exclusion*. Bristol, UK: Policy Press.

Le Grand, J. 1998. Possible definition of social exclusion. In *CASE meeting*. Cumberland Lodge: London: CASE.

Lloyd-Jones, T., and A. Brown. 2002. Spatial planning access and infrastructure. In *Urban livelihoods—A people-centred approach to reducing poverty*. London: Earthscan.

McKay, L. 1987. Tourism and changing attitudes to land in Negril, Jamaica. In *Land and development in the Caribbean*. J. Besson, ed. London: Macmillan.

Mvungi, A., and M. Ndanshau. 1999. The Congo street culture: Creation of wealth and employment through petty trade. Dar es Salaam, Tanzania: National Housing Development Corporation.

Nnkya, T. 2003. *A rights-based approach: Governance, context and experience of public space and livelihoods in Dar es Salaam. Preliminary Research Report*. Cardiff, UK: Cardiff School of City and Regional Planning.

Oppermann, M. 1993. Tourism space in developing countries. *Annals of Tourism Research* 20:535–56.

Pattullo, P. 1996. *Last resorts: The cost of tourism in the Caribbean*. Kingston: Ian Randle Publications.

Population Census 2001, Jamaica—Country Report. 2001. Kingston: Statistical Institute of Jamaica

Robson, C. 1993. *Real World Research: A Resource for Social Scientists and Practitioner-Researchers*. Oxford: Blackwell Publishers, Ltd.

Sibley, D. 1995. *Geographies of exclusion: Societies and difference in the West*. New York: Routledge.

Sindiga, I. 1996. International tourism in Kenya and the marginalisation of the Waswahili. *Tourism Management* 17 (6):425–32.

Social Exclusion Unit. 1997. Bringing Britain together: A national strategy for neighbourhood renewal. London: The Cabinet Office.

UN HABITAT 2003. Slums of the World: The face of urban poverty in the new millennium. Nairobi: United Nations Human Settlement Programme.

Ulack, R. 1978. The role of urban squatter settlements. *Annals of the Association of American Geographers* 68 (4).

Van Dierman, P. 1997. *Small business in Indonesia*. Aldershot, UK: Ashgate.

Verschoor, G. 1992. Identity, networks and space: New dimensions in the study of small-scale enterprise commoditization In *Battlefields of knowledge: The interlocking of theory and practice in social research and development*. I. Long, ed. London: Routledge.

10 An Unwelcome Guest
Unpacking the Tourism and HIV/ AIDS Dilemma in the Caribbean: A Case Study of Grenada

Wendy C. Grenade

INTRODUCTION

In the lead up to the 2007 Cricket World Cup (CWC),[1] tourism officials in the Caribbean anticipated a boost in tourism, given the expected influx of visitors to the region. As the most tourist-dependent region in the world,[2] officials were anxious to capitalize on this opportunity to showcase the Caribbean and at the same time benefit from tourist expenditures. Governments invested heavily in capital projects to accommodate the games. The construction sector boomed as stadia were erected and road networks built or refurbished. Yet despite the fact that CWC was expected to generate economic benefits, there were concerns about an unwelcome guest—HIV/ AIDS—which unfortunately is no stranger to the Caribbean. This region has the second highest adult prevalence rate after Sub-Saharan Africa.[3] In 2006 there were 250,000 people living with HIV in the Caribbean. Of that number, nearly three-quarters are in Haiti and the Dominican Republic. However, national adult HIV prevalence is high throughout the region: 1 percent to 2 percent in Barbados, Dominican Republic, and Jamaica and 2 percent to 4 percent in the Bahamas, Haiti, and Trinidad and Tobago. Cuba is the exception, with prevalence below 0.1 percent. In 2000 the United Nations secretary-general noted that HIV/AIDS is a global problem, not just an African one, and that it threatens human security worldwide. As a consequence, in 2002 the United Nations Security Council passed Resolution 1308, which indicated that HIV poses a security threat to the nations of the world.

Within this context, policymakers in tourist-dependent economies face a predicament, since tourism may be a vehicle which helps to spread diseases, such as HIV/AIDS. At the same time, high HIV/AIDS prevalence rates could undermine the tourism industry and have negative implications for socioeconomic development. Hence, '[t]he islands with the tourism are the ones that don't talk about it very much.'[4] Events such as the recently concluded CWC bring to the fore the policy dilemma which surrounds tourism dependency and the threat which HIV/AIDS poses for development. This chapter uses the case of Grenada to probe the problem. The purpose of the chapter is to contribute to the debate on the challenges which small developing states face in this era of globalisation and in particular the contradictions inherent in tourism development.

This work is relevant and timely for several reasons. As regional integration deepens, through the Caribbean Single Market (CSM),[5] opportunities may be created for greater intra-Caribbean travel for business and pleasure. This expected increased mobility, while necessary to promote economic activity, has the potential to increase the spread of the epidemic.[6] Further, the influx of visitors for the 2007 CWC triggered concerns about sex tourism. Finally, in the special case of Grenada, in 2004 Hurricane Ivan wreaked havoc on the island, creating severe damage to the tourism sector. As the country seeks to rebuild its economy and social infrastructure, this study is intended to be a useful policy tool. This is by no means an exhaustive study. However, it seeks to contribute to the continued search for answers to the challenges that bedevil the Caribbean.

The chapter begins with a general overview of the interrelationship between tourism and HIV/AIDS in the Caribbean. It then goes on to examine the special case of Grenada, using a four-pronged typology: economic, environmental, social, and institutional vulnerability (Thomas 2004). It concludes by considering strategies for the way forward. The data for the study is based on scholarly articles, policy documents, and a series of interviews which were conducted in Grenada during the period 2003–06.

THE INTERRELATIONSHIP BETWEEN TOURISM AND HIV/AIDS IN THE CARIBBEAN: A GENERAL OVERVIEW

The interrelationship between tourism and the spread of diseases has its roots in the historical development of the Caribbean. The Caribbean has had a long legacy of sexual exploitation. Reference to sex work on studies of slavery in the region suggests that it was an integral part of the region's history (Beckles 1989; Morrissey 1989). In the postcolonial era there continues to be power differentials between tourists and locals as well as constructed images about Caribbean sexuality. The region's peoples are presented in tourist imaginations in stereotypical roles such as 'hypersexual black male stud' and the 'hot mulatta or black woman' (Kempadoo 1999). The destination is portrayed as 'feminine, yielding, powerless and vulnerable, with the woman represented as exoticised commodities, which are there to be experienced' (Pritchard and Morgan 2000:891, cited in Barrow 2003). According to Boxill et al. (2005:23), 'the tourist industry is founded on the idea of providing a place free from normal social constraints, a relaxed, often times hedonistic atmosphere, where consequences do not exist. For the tourist, it serves to satisfy those desires that are "forbidden fruit" at home.' As Allen and Nurse (2004:17) note, within this context, it is not surprising that tourism-dependent economies have some of the highest HIV prevalence rates and reported AIDS incidences in the region. For example, as pointed out above, HIV prevalence rates are relatively high in the Bahamas, Barbados, the Dominican Republic, and Jamaica, which are key tourist destinations. Camara (2001) notes that the initial problem of the HIV/AIDS epidemic in the Caribbean is traced to gay sex tourism

Across the region sex tourism or tourism-oriented prostitution has become an important topic of research and discussion due to the growing reliance of national governments on income generated by tourism and tourism-related activities. Karch and Dann (1981) did research on 'beach boys' in Barbados, drawing attention to the negotiations that take place between black Barbadian men and white women around their sexual, gendered, and racialised identities as well as the way in which the relationships were shaped by the location of Barbados as a Third World nation locked into dependency in the global economy. The Dominican Republic, Cuba, and Jamaica have also been sites for research on tourism-related prostitution (Barrow 2003).

Several Caribbean governments and civil society groups are concerned about the impact of sex tourism. Officials in Trinidad and Tobago have expressed concern about the growth of sex tourism, the so-called beach-bum phenomenon, and the link to the spread of HIV.[7] In the Dominican Republic, AIDS activists are concerned about child prostitution in resort areas and the spread of HIV.[8] In a study conducted on HIV/AIDS and tourism in Jamaica and the Bahamas, it was found that 'a significant number of visitors said that they used the trip as an opportunity to find new sexual partners locally' (Boxill et al. 2005:21).

Evidence suggests that sex tourism and the spread of HIV/AIDS are directly related to poverty. As American Foundation of AIDS Research[9] points out, '[p]overty and economic disparity are forcing men and women into commercial sex work, often with tourists, [and that] many of the same factors contributing to the quick spread of HIV through Sub-Saharan Africa—extreme poverty, malnutrition, poor health care, and high rates of migration—also afflict the Caribbean and Latin America.'[10] Governments have begun to take notice of this apparent connection and have called for further study. At the International Migration Policy seminar held in Jamaica in May 2001, government representatives noted the need to 'further identify the link between HIV/AIDS and migration, particularly tourism, business travel and internal migration' (Borland, R. et al 2004). As the Caribbean Community (CARICOM) confirms:

> Commercial sex work is widespread, well entrenched and increasing throughout the region, and takes place under a variety of circumstances. It is linked to tourism in the islands; it follows mining villages and trading patterns in a variety of industries. There are short-term as well as fixed brothel workers, and mobile sex workers; they are single

and married, women and men. Male prostitution in the form of 'beach boys' is increasing across the Caribbean. In many cases economic hardship is the single most important reason given by sex workers for going into sex work. Economic difficulties in the region and the rigors of structural adjustment over the last two and a half decades have resulted in a dramatic rise in the number of women and men seeking work in a market that is less than accommodating (CARICOM 2000:7).

This problem of tourism and HIV/AIDS reflects the contradictions inherent in the new global order. On the one hand, while globalization provides many opportunities (Ohmae 1990; Naisbit 1994), it also has negative consequences, particularly for the developing world (Wood 2000). In the case of the Caribbean, this region has had to adapt to many rounds of it (Klak 1998). For example, in the postcolonial period, most Caribbean countries underwent Structural Adjustment Programmes (SAPs) imposed by the International Monetary Fund (IMF) and the World Bank (WB). Those austerity measures had a crippling effect on already fragile economies of the region. Some of the major consequences were:

- The 'streamlining' of the public service, which helped to further weaken institutional capacity;
- High unemployment;
- Increased social ills, such as illicit drugs, crime, violence, and HIV/AIDS;
- The perpetuation of poverty; and
- Escalating debt.

On the question of debt, the WB notes that the Organisation of Eastern Caribbean States (OECS) has the 'dubious distinction of hosting six of the world's most highly indebted emerging economies' (World Bank 2005:vi). The ensuing debt burden has severe implications for all sectors of the economy. To service the debt requires the reallocation of scarce funds from sectors such as education, health, and tourism.

At the same time, globalisation and trade liberalization have created a highly competitive global trade environment in which small, postcolonial economies are quite vulnerable. For instance, one of the landmark cases was the banana dispute[11] between the European Union (EU) and the United States and the subsequent World Trade Organisation (WTO) ruling. This heralded the end of preferences and had severe implications for the banana-producing countries in the CARICOM region. Therefore, with the end of preferences and falling world prices for commodity goods, Caribbean economies diversified heavily into services such as offshore financial services and tourism. However, while the services sector provides opportunities to capitalize on the restructuring of the global political economy, it has its own vulnerabilities.

Policy Responses

National responses to the problem of HIV/AIDS and tourism have been varied. On one extreme, the Cuban government frontally addressed the issue through a controversial programme. It instituted mandatory testing in the early stages of the epidemic, which required all HIV-positive individuals to live in government-run sanitaria to protect the collective rights of citizens and preserve the tourism sector. Consequently, as noted above, Cuba has one of the lowest HIV prevalence rates in the Caribbean.

Thailand is another tourism-dependent economy that has had to grapple with the problem. As a country hard hit by the epidemic, Thailand has been cited by United Nations agencies, the WB, and other organizations as having conducted one of the world's most successful AIDS prevention and awareness programmes. A couple of the key lessons learnt include:

- The leadership must recognize the devastating scale of the epidemic and be willing to discuss openly the enormity of HIV/AIDS;
- Policymakers need to fight HIV/AIDS from both the preventive and treatment perspectives;
- HIV/AIDS needs to be addressed in a holistic manner—addressing cultural, economic, human, and social aspects of the epidemic;
- All sectors of the society—from the highest to the grass-root levels, including nongovernmental organizations and people living with HIV/AIDS—should unite to tackle the pandemic.[12]

Other countries have opted for a more cautious approach. For example, the Barbados government enunciated a new approach to AIDS awareness and protection which included a focus on the tourism sector. According to a government official, the minister of state in the prime minister's office, 'Government acknowledges the strong link between tourism, sex workers and the HIV epidemic and appreciates the pressing need to protect an industry that is the economic mainstay of Barbados and at the same time protect the people of Barbados from associated risks.'[13] In 2002 the Barbados government launched what was termed a 'warning welcome' to visitors. Tourists arriving at the airport were presented with brochures which featured an HIV/AIDS awareness message from Prime Minister Owen Arthur. The initial response from a few visitors ranged from fear to offence. Following this, concerns were raised about the appropriateness of this initiative and the reaction from visitors[14] (Barrow, 2003). The warning welcome had to be discontinued.

This highlights the conflict between the HIV/AIDS prevention community and the tourism industry. For instance, a former chief executive officer of the Caribbean Hotel Association, cautioned against making 'a big issue of the region's AIDS epidemic.'[15] He noted that regional private-sector interests tend to view sex tourism as a very profitable niche and observed that resorts which market sex more explicitly, such as Hedonism in Jamaica and Club

Med, tend to have high occupancy rates. Therefore, in the official's view 'sex tourism is simply a matter of using what sells.'[16] As he further stated: 'We as an industry will always use the things that are most attractive, most exciting . . . [a]nd what's more attractive and exciting than sex?'[17] (cited in Barrow 2003: 177–81). Herein is the problem. HIV/AIDS can be sexually transmitted and sex tourism is a profitable niche in the highly competitive tourism market. This is a grave policy dilemma.

A Regional Policy Response

Policy makers in CARICOM have devised a regional response to the HIV/AIDS problem. In 1998 the Caribbean Task Force on HIV/AIDS was formed. A Caribbean Regional Strategic Plan of Action for HIV/AIDS was launched in 1999–2004. In 2001 the Pan Caribbean Partnership Against HIV/AIDS (PANCAP) was established. PANCAP has 61 signatories and includes country governments, people living with the disease, intergovernmental organizations, private sector entities, religious groups, and the academic community. It aims to build awareness, provide information, and monitor programmes. Areas of focus include mother-to-child transmission, capacity building to manage and care for people infected with the disease, and communications. In 2002 an Accelerated Access Initiative Task Force was formed. In 2003 there was the regional application of the Global Fund. Prime Minister Denzil Douglas of St. Kitts/Nevis has lead responsibility in CARICOM for human resource development and HIV/AIDS, and he has been visibly at the forefront in the fight against HIV/AIDS, lobbying for assistance from the Clinton Foundation, for example. A number of international institutions have given assistance to the Caribbean. For example, the WB has approved loans to Barbados (2001), the Dominican Republic (2001), Jamaica (2002), Grenada (2002), St. Kitts and Nevis (2003), Trinidad and Tobago (2003), PANCAP (2004), Guyana (2004), St. Lucia (2004), and St. Vincent and the Grenadines (2004; Sullivan 2006:2).

While the regional response to HIV/AIDS is commendable, this chapter argues that there is an urgency for policymakers to mount a similar strategy to address the question of HIV/AIDS and tourism in the Caribbean. This is particularly necessary in light of the recent launch of the CSM and the expected increased movement of people throughout the region. Investigations reveal that there is no clear regional policy statement which holistically addresses HIV/AIDS and tourism.[18] Within this context, the following section examines the problem by focusing specifically on Grenada.

THE PUBLIC POLICY DILEMMA: TOURISM AND HIV/AIDS IN GRENADA

The tri-island state of Grenada, Carriacou, and Petite Martinique is the most southerly of the Windward Islands. It has a land area of 345 sq km (133 sq miles) and a population of 106,500 (2005).[19] The population structure is young, with 31.4 percent of the population below the age of 15 years, and 10.32 percent are 65 years and over. The 1991 census showed ethnic composition of the population as black (85 percent), mixed (11 percent), and East Indian (3 percent). Grenada is a former British colony which obtained political independence in 1974. As a relatively newly emerging economy, Grenada is searching for a path to development in the context of a troubled political history; a hostile global political economy; environmental vulnerability; weak human, financial, and institutional capacity; and nontraditional security threats, such as illicit drugs, crime, violence, and HIV/AIDS.

Politically, Grenada experimented with a socialist revolution during the period 1979–83 and was invaded by the United States in 1983. Grenada is currently a parliamentary democracy. However, it has been described as 'an island of conflict' (Brizan 1984). Postrevolutionary politics in Grenada is described as 'chaotic, confusion-making, very sectarian and very tribal.'[20] This chapter argues that the socioeconomic condition of Grenada has to be understood within the context of the political development of the country. This section uses the case of Grenada to explain the public-policy dilemma which surrounds tourism and HIV/AIDS in the Caribbean.

To conceptualise the problem, it is useful to engage the discourse on the vulnerability of small states in the global environment.[21] The central argument is that small states are relatively more vulnerable to shocks, given historical, structural, and geographical conditions. This vulnerability in turn undermines development efforts. Thomas (2004) captures the essence of the debate in his four-prong vulnerability typology. Thomas (2004:1) defines vulnerability as 'proneness of small states to the adverse effects of changes in their environment and insufficient resilience to overcome these adverse affects on their own.' Thomas notes that the definition twins 'a limited capacity to respond' together with 'exposure to shocks' and is directed at four major dimensions of small states: economy, environment, society, and the institutional framework. Each will be looked at in turn.

First, Thomas (2004:1–3) defines economic vulnerability as 'the greater than average risk small economies face from exogenous shocks which adversely affect their incomes, employment, output, markets, consumption and wealth.' Second, environmental vulnerability is defined as 'the greater than average risk small economies face of damage to their natural ecosystems.' Third, Thomas refers to social vulnerability as 'the greater than average risk posed by internal and external factors in undermining social cohesion, introducing systematic pathologies and eroding social capital.' He cites examples such as illicit drugs, violence, organized corruption, and HIV/AIDS. Fourth, institutional vulnerability refers to 'the greater than average risk posed by the limited capacity of domestic institutions to respond to the complexity and intensity of the pressures flowing from globalization.' This chapter uses Thomas's framework to explain the problem of tourism dependency and the spread of HIV/AIDS in Grenada.

Economic Vulnerability

As is the case with many other small developing states, Grenada is economically vulnerable. It is one of the eight island states which comprise the Eastern Caribbean Currency Union (ECCU)[22] and a member of the CSM. Table 10.1 provides selected statistics of the ECCU. As Table 10.1 shows, Grenada's GDP per capita (2005) was US$3,260, compared to US$7,983 in Antigua and Barbuda; US$5,660 in St. Kitts and Nevis and US$3,798 in St. Lucia. For the period 1980 to 2003, Grenada's relative growth per capita was 2.8 percent, the lowest in the ECCU. However, inflation has remained relatively low in Grenada. The inflation rate has averaged 2 percent over the last 15 years. As is the case with other ECCU countries, Grenada is faced with high public debt. By the end of 2003 public debt spiralled to 110 percent of GDP (IMF 2005). However, following a contraction in economic activity of 0.4 percent in 2002, the economy recovered and registered a positive growth rate of 5.7 percent in 2003 fuelled by expansion in the hotels and restaurants, transportation, construction, and wholesale and retail trade sectors. In particular, the robust performance of the tourism industry was an indication that the sector was recovering from the effects of September 11, 2001.[23]

As a small island with an open economy, Grenada is vulnerable to external shocks. Thomas (2004:1–2) defines economic vulnerability as the 'the greater than average risk small economies face from exogenous shocks which adversely affect their incomes, employment, output, markets, consumption

Table 10.1 Selected Statistics ECCU

	Pop ('000s) 2003	GDP PC (US$m) 2005	GNI PC (US$) 2005	Growth pc (%) 1980- 2003	Poverty (% pop) latest	Unemp (% LF) Latest
Antigua & Barbuda	79	7,983	7,619	4.2	12	7
Dominica	71	7,891.51	2,615	2.9	33	25
Grenada	105	8,804.55	2,919	2.8	32	13
St. Kitts & Nevis	47	15,300.68	5,135	4.7	31	5
St. Lucia	161	10,255.16	3,469	3.2	25	19
St. Vincent & The Grenadines	109	8,077.64	2,828	3.5	38	21

Source: World Bank 2005 and Eastern Caribbean Central Bank, 2005

and wealth.' Agriculture was traditionally the mainstay of Grenada's economy with nutmegs, cocoa, and bananas as the main export crops. Over the last two decades the economy has shifted from the once-dominant agricultural sector into services, which now account for over three-quarters of total value added (IMF 2005). Since most of the goods consumed in the country are imported, Grenada is susceptible to changes in global economic conditions. Additionally, since Grenada is heavily dependent on the tourism sector, global forces impact on Grenada's economy. For instance, according to the IMF, the September 2001 terrorist attacks and the slowdown of the global economy had a negative effect on Grenada's economy. Real GDP growth, on average, has been low over the last five years compared with growth of nearly 6 percent a year in the late-1990s. Therefore, Grenada is economically vulnerable because of its colonial legacy of dependency, its open economy, and its location in the global economy.

Grenada's Tourism Sector

The tourism sector is an important sector in Grenada's diversification strategy. It is a significant source of foreign exchange and employment generation. The development of tourism has also helped to cushion the effects of the decline in its exports, particularly bananas and cocoa. Table 10.2 depicts sectoral contribution to GDP for the period 2002–05.

Table 10.2   Sectoral Contribution to GDP—Grenada 2002–05

SECTOR	2002	2003	2004	2005
Agriculture	65.70	64.10	59.40	37.98
Manufacturing	54.32	52.99	45.23	40.93
Construction	53.59	67.53	66.96	87.05
Hotels & Restaurant1	54.09	61.58	53.53	48.18
Transport	88.23	94.82	99.75	103.34
Communication	84.00	85.58	91.74	94.95
Banks & Insurance	78.89	85.20	87.33	89.95
Government Services	93.68	94.23	97.54	99.49
Total	676.30	715.55	693.91	707.11
Growth Rate	0.84	5.80	-3.02	1.9

Source: "Grenada National Plan for Health 2006-2010"
Situation Analysis, October 2005

1. Hotel and Restaurant represent the tourism sector

As Table 10.3 shows, visitor expenditure increased from EC$155,527,754 in 1995 to EC$417,763,644 in 2004. This rise in tourism activity can be attributed to the expansion of the country's hotel capacity and the upgrading of its tourist facility during the 1990s. The number of rooms in tourist accommodation establishment rose from 1,115 in 1990 to 1,758 in 2003. For the first two quarters of 2006, the number of stay-over visitors increased by 37.5 percent to 33,280, associated with the reopening of hotels that were closed in the first quarter of 2005. Increases were recorded in arrivals from the three major markets—Canada (92.5 percent), Europe (85.7 percent), and the United States (27.1 percent) (Grenada Board of Tourism 2004: 60–61).

The number of visitor arrivals dipped during the period 2001–02. This can be attributed to the September 11, 2001, attacks in the United States of America and Grenada's vulnerability to external shocks. Nonetheless, according to the Organisation of Eastern Caribbean States (2004), the contribution of tourism to the economy grew from 5.8 percent in 1990 to 9 percent in 2000 and has remained roughly at that level.

Thus, tourism is a lead growth sector in Grenada's economy and, despite limited financial resources, the government of Grenada (GOG) still attempts to allocate relatively adequate amounts to this sector. Commencing in 2000 the government committed EC$10 million[24] per year for three years to the Grenada Board of Tourism (BOT). The money was earmarked for strategic marketing programmes with airlines; destination marketing activities; product enlargement and environment management; small hotels specialized assistance programme, and human-resource development (Government

Table 10.3   Visitor Arrivals and Expenditure, Grenada—1995–2004

Year	Visitor Arrivals(#)	Visitor Expenditure(EC$)
1995	369,336	155,527,754
1996	386,013	160,350,146
1997	368,417	398,640,834
1998	391,680	415,314,488
1999	378,952	457,562,230
2000	316,528	480,465,614
2001	277,557	438,938,320
2002	271,394	468,476,074
2003	294,211	469,514,388
2004	369,810	417,763,644

Source: Grenada Board of Tourism, Annual Statistical Report, 2004: 60–61.

of Grenada 1999:27–28). The 2005 budget provided EC$9 million for the marketing and promotion of the destination and a further EC$2 million for joint marketing and risk-sharing agreements with selected airlines. In terms of tourism marketing, the GOG increased promotion from EC$9 million in 2005 to EC$12 million in 2006. The Government of Grenada invested in a modern cruise-ship terminal in 2004. Subsequently, the number of cruise-ship visitors increased from 146,925 in 2003 to 222,944 in 2004, an increase of 54.5 percent (Government of Grenada 2005). Cruise tourism provides benefits to a wide range of people including taxi drivers, water taxis, tour operators, and restaurants. Thus, tourism can be viewed as one of the bed-rocks of the Grenadian economy. Yet tourism is a volatile industry which can affect and be affected by global forces. There are no explicit policies which link tourism and HIV/AIDS. Yet the GOG acknowledges that HIV is an economic risk.

According to Prime Minister Mitchell, '[w]e are faced with a danger-ous risk—a risk that we cannot afford financially or socially.'[25] The prime minister further pointed out that cost is one of the impediments which prevents government from effectively tackling the problem. According to Prime Minister Mitchell, '[t]remendous resources are required, which are beyond the reach of government.' In the 2001 budget speech, the minister of finance confirmed that the cost of medical care for HIV/AIDS patients is 'unbelievably high' and will place a financial burden on both the families of the patients as well as the government (Government of Grenada 2001:16). Officials in Grenada are concerned about the long-term effects of HIV/AIDS on the economy. According to the former health minister:

> The threat is very serious especially since the age group most vulnerable is 15–25, which is our most productive group. These are our youths to take up leadership roles in our country, in our workforce, in our society as a whole. They carry the burden of the development of our country. If we lose those persons we lose the very future of our country. This is a serious economic and security risk. The cost of treating HIV/AIDS is high. Monies that could have otherwise been utilized for the develop-ment of the country are now being diverted to treat persons living with the disease. If we did not have HIV/AIDS we could have been using those funds elsewhere—for health, education or tourism. However, as difficult as it is, any money we spend on the disease now is an invest-ment for the country. We have to see the long-term economic threat—it can make the country even poorer.[26]

Therefore, one of the public policy dilemmas which surrounds tourism and HIV/AIDS relates to economic vulnerability of the country. On the one hand, the economy is open and susceptible to external shocks. At the same time, there is a high debt burden, which is one of the consequences of SAPs. Concurrently the country is dependent on tourism, which is a volatile industry. With scarce financial and human resources, government was already burdened. Given the reality of HIV/AIDS, there is an added burden to find resources to fight the epidemic. One avenue is the tourism sector, but tourism itself may be a vehicle for the spread of the disease. At the same time, as funds are diverted to combat HIV/AIDS, this takes away resources needed to market and promote sustainable tourism. This is a dilemma for policymakers.

Environmental Vulnerability

As is the case with other ECCU countries, Grenada is also vulnerable to environmental shocks. Thomas (2004:1–2) defines environmental vulnerability as 'the greater than average risk small economies face of damage to their natural eco-systems.' In Grenada's case, in 1999 tidal surges left a trail of destruction to the physical and economic infrastructure of the country. Damages were estimated at approximately US$100 million, that is, one third of GDP in 1998. Consequently, additional resources had to be mobilized for reconstruction purposes (Government of Grenada 2001:2). In 2004 Hurricane Ivan severely destroyed Grenada. The damage from the hurricane exceeded 200 percent of GDP (EC$2.4 billion). Following are some of the impacts:

- Virtually every house on the island was affected, with 30 percent completely destroyed; schools and universities suffered extensive damage, as did hospitals and shelters;
- Productive sectors were hit hard. In the tourism sector, few hotels escaped damage [the hurricane affected close to 90 percent of the tourist accommodations].[27] In agriculture, the nutmeg sector—in which Grenada was the world's second largest producer and which employed an estimated 30 percent of the population directly or indirectly—was decimated;
- The entire population was without water and electricity in the immediate aftermath of the hurricane;
- Nearly 8 percent of the labour force was displaced from their jobs in the immediate aftermath of Ivan, raising the unemployment rate to over 20 percent. Starting from an already high incidence of over 30 percent, poverty levels have reportedly risen sharply among farmers and women (IMF 2005; OECS 2004).

The IMF reported that prior to Hurricane Ivan, the GOG was making progress to restore fiscal sustainability and spur growth. The primary fiscal balance had registered a surplus in 2003 for the first time in nearly a decade and was projected to remain in surplus in 2004. Real GDP was projected to grow by over 4 percent in 2004. However, this situation was set back by Hurricane Ivan. The severe impact of the hurricane forced the government to change its economic priorities. It also compounded the tourism-HIV/AIDS problem. For example, as part of the reconstruction efforts, an influx of soldiers, humanitarian workers, representatives from donor agencies, Grenadians living in the diaspora, and other groups visited Grenada. As is the case in complex emergencies, the immediate post-Ivan period saw, for example, reports of relief supplies being used by some members of the security forces in what was described as a lucrative 'food for sex in Grenada.'[28] This no doubt may have added to the tourism-HIV/ AIDS challenge.

Thus, environmental vulnerability complicates the already fragile socio-economic condition of small countries. Besides the social ills that accompany the immediate postdisaster period, resources have to be reallocated from key sectors to facilitate rebuilding efforts. Therefore, the severity of Hurricane Ivan shows Grenada's vulnerability to natural disasters which can further undermine socioeconomic development.

Social Vulnerability

Thomas (2004:3) refers to social vulnerability as 'the greater than average risk posed by internal and external factors in undermining social cohesion, introducing systematic pathologies and eroding social capital.' He cites examples such as illicit drugs, violence, organized corruption, and HIV/ AIDS as factors which give rise to social vulnerability. Social vulnerability is a major problem which faces Grenada. According to the IMF (2005), Grenada's social indicators are weaker than in other ECCU countries. For example, out of 177 countries, Grenada's human development index (2004) was ninety-three, while the average in the ECCU was seventy-three. Table 10.4 shows the HDI for the ECCU for the years 2004–06.

The IMF also pointed out that Grenada's GDP per capita income (2004) was US$4,205, while it was US$5,473 in the ECCU. Life expectancy at birth

Table 10.4   HDI in ECCU 2004–06

Country	HDI2004	HDI2005	HDI2006
Antigua & Barbuda	55	60	59
Dominica	95	70	68
Grenada	93	66	85
St. Kitts & Nevis	39	49	51
St. Lucia	71	76	71
St. Vincent & The Grenadines	87	87	88

Source: United Nations Development Programme, 2002–2005.

(2003) was seventy-three in Grenada, while the ECCU average was seventy-four. Infant mortality rate per one thousand births (2003) was eighteen in Grenada while the ECCU average was seventeen. Grenada's adult illiteracy rate (percent) (2001) was 6, whereas it was 8 in the ECCU. Grenada's poverty headcount index (2000) was thirty-two, whereas it was twenty-nine in the ECCU (IMF 2005: 8). These indicators reflect the level of social vulnerability in the country. The following section specifically discusses the HIV/AIDS as a major threat to social cohesion in Grenada.

HIV/AIDS IN GRENADA

In 1984 the first case of HIV was diagnosed in Grenada. Since then, 223 persons have been diagnosed with HIV (See Figure 10.1). Of that amount, 58.9 percent have developed AIDS while 82.95 percent of those diagnosed with AIDS have subsequently died. The prevalence rate of the disease has been estimated at <0.5 percent.

As Figure 10.1 shows, within the past five years incidences of the epidemic are increasing. The former minister of health indicated that the statistics 'are alarming' because Grenada is a small country with just over one hundred thousand, with over two hundred persons already infected, based on reported cases. The minister refers to the epidemic as 'a time bomb that is ticking and if it is not diffused it has the potential to seriously explode.'[29]

Since the first case of HIV was diagnosed in 1984, Grenada embarked on a prevention campaign to fight the deadly virus. Judy Benoit, an officer in the Ministry of Health (MOH,) has been at the forefront of the fight since

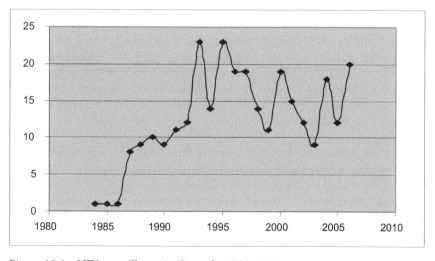

Figure 10.1 HIV surveillance in Grenada 1984–2006.

1984. According to Benoit, '[i]n the past we focused on HIV prevention campaigns since we were seeking to bring about behavioural change. The first mission was to inform the nation about prevention. We have long gone past that prevention stage now.'[30] The prevention messages have been consistent and strong: ABC—Abstinence, be faithful, use a condom. However, despite years of vigorous prevention campaigns, the epidemic is still increasing. The minister for education lamented that:

> For years we have heard the prevention message. The MOH has said and done a lot. But we still see the rise in the incidences of HIV. The difficulty is, sex is an emotional thing, it is a private thing and the problem is more difficult that we think. We have tried everything from giving information in schools to distributing condoms. The reality is, we cannot legislate or implement policies that can guarantee a change in sexual behaviour. We have to understand that behaviour is influenced by our environment, socio-economic conditions, traditions and culture.[31]

Consequently, the strategy had to be modified to take into account the fact that although financial and human resources were being allocated to prevention campaigns, this was not resulting in behavioural change. As Benoit further pointed out:

> This [2006] is almost 25 years after the start of the epidemic and we are still seeing an increase here in Grenada. For the whole of last year we identified 12 new cases, by the end of March this year we had 15, by the end of August we already have 20 and the year is not yet ended. With all the information out there, with all the care and treatment, and with voluntary counselling and testing HIV/AIDS is still on the increase. We are at a stage now where we are looking at various factors that drive the epidemic, such as poverty, unemployment, deviant youth, gender inequity, gender inequality, religious norms, among others. In terms of poverty, there are parents who are struggling to send their children to school and some of them will do anything, including commercial sex work, to get that dollar to buy the milk so that the children could go to school with something on their stomach. We have done a lot in the past but we have been slow to change our strategies to reflect today's realities.[32]

The data support the fact that there are many factors responsible for the spread of the disease. According to the 2003 *Strategic Plan for HIV/AIDS in Grenada,* one of the factors favourable to the spread of the disease is the 'existence and acceptance of transactional/commercial sex, especially within the context of the significant poverty levels that exist in the country.' The 1999 poverty survey estimated that 32.1 percent of all individuals in Grenada were poor. That is, their annual expenditure was less than EC$3,262, which is the cost to meet their minimal food and other requirements. Of all individuals, 12.9 percent were found to be extremely poor or indigent. The report also pointed out that poverty particularly affected youth, with over 56 percent of the poor being less than 25 years old (Poverty Assessment Report, Grenada 1999). The former minister for health confirmed that 'Poverty is a culprit in all of this. Poverty goes with lack of knowledge, information and education. Commercial sex is not practiced overtly in Grenada, but it does happen. Poverty makes people more vulnerable and they often turn to alternative survival strategies.'[33]

The minister for finance also indicated that there are still some people in Grenada who go to bed hungry not knowing where their next meal is coming from, or how they will send their children to school because they have no food, or cannot afford to buy books or uniforms for their children (GOG 2001:23). Sandra Ferguson, a leading nongovernmental organisation (NGO) representative, pointed out that from her work in rural communities, Grenada has not yet recovered from SAPs and there are widening poverty gaps. 'We are seeing the new working poor,' she asserted.[34] The director of the National AIDS Directorate also confirmed that '[p]overty is one of the leading factors driving the epidemic in Grenada since the population group most affected by HIV/AIDS are made up of low income earners.'[35] According to the minister for tourism, Grenada recognizes the importance of tourism to poverty alleviation, and if structures are not put in place to address poverty, the destination can lose the very tourism on which it depends. This in turn can have a negative impact on the economy as a whole.[36]

Gender issues also contribute to the complexity of the problem. The statistics show that a greater percentage of women are affected by the HIV/AIDS epidemic. In many cases, there is the problem of 'sex for security.'[37] This has implications for the interrelationship between tourism and HIV/AIDS. Benoit confirmed that Grenada's first case of HIV/AIDS came from abroad. While some tourism officials argue that sex tourism is not a problem in Grenada, as is the case in some other Caribbean destinations, the director of the National AIDS Directorate pointed out:

> We all know that visitors come to the Caribbean looking for sex. Let's not fool ourselves. And how often would tourists pack condoms in their suitcases? We have three serious issues here. On the one hand, visitors to our islands come to have fun and the values they usually uphold in their home countries they do not often adhere to in a foreign country, particularly when they are in a holiday mode. They may have sex as part of fun without thinking they are multi-partnering. They drink extra alcohol, use harder drugs; no one sees them, so it's ok. Second, once the tourist is coming with foreign currency in a poor country there will be a thriving commercial sex trade. Third, Caribbean people are generally friendly and welcoming. Since we depend on tourism we often feel obligated to please the tourists—giving them the whole experience. Therefore, many factors contribute to the complexity of the problem.[38]

However, there is little or no research done on sex tourism in Grenada and commercial sex is illegal (Criminal Code Chapter 1 Volume 1 of the 1994 Continuous Revised Laws of Grenada [CAP 126]). In the run-up to CWC 2007, there was a debate about the handling of commercial sex workers. At a two-day CARICOM/CWC 2007 Health Sector Sub-Committee meeting in August 2006, Antigua's health minister tabled the idea of legalizing prostitution during the 51-day tournament. He pointed to the likely influx of commercial sex workers (CSWs) in the eight Caribbean countries that will host the games. 'By registering and regulating the women [and men] the spread of HIV/AIDS can be curtailed,' he argued. Barbados and St. Kitts and Nevis said no to this proposal.[39] According to Grenada's minister for health:

> Grenada has launched the HIP (How you Playing Program) since it is expected that there will be an influx of visitors around world cup—we are ensuring that we continue to advocate the ABC—abstinence, faithfulness and condom use—for our nationals and tourists. Grenada is predominantly a Christian society—commercial sex is not an authorized means of earning a living. And we have no intention as a government or indeed as a people to legalise commercial sex work, prostitution or even homosexuality. But we ensure that we treat everyone with dignity.[40]

As the minister's position suggests, cultural and religious norms are key factors in the milieu. The 1991 census showed religious affiliations as Roman Catholic 53.1 percent, Anglican 13.9 percent, Seventh-day Adventist 8.6 percent, and Pentecostal 7.2 percent.[41] There is, for instance, an issue with the Catholic Church and its strong stance against the distribution of condoms. The minister of education asked, 'How do you ask a Catholic to wear a condom and how will the church react if the government promotes condom use?'[42] Regarding the role of the church, Prime Minister Mitchell indicated that 'The church must continue to use the pulpit to reach out and sensitise the public. But we have to go further. We tell people do not commit sin, but they *do* commit sin. In cases where they disobey, the state has a responsibility to protect the collective.'[43]

This adds another dimension to the problem. That is, in a democracy, how does one balance the need to promote human rights and protect national security? Security in the Grenadian context relates to the 'capacity of the system to sustain life.'[44] In relation to the Cuban model, Prime Minister Mitchell indicated that such an approach will be very difficult to implement in a democratic society. However, he stated that Grenada can find the balance through legislation. According to the former minister for health, 'mandatory testing is very difficult. When you start interfering with human rights it is a very serious issue.'[45] According to the current minister for health, 'At times the public health issue should be first and foremost. In this case the public health issue should come first.' In her view, there is need to protect the whole, even the person 'who is kicking up against it.'[46] The minister for education also shared this view. In relation to the Cuban model, she pointed out that 'this is the best thing we can do [for] if you don't have mandatory testing the Government will not be able to take care of the budget in such a way for the very people who would need care and treatment.'[47] Therefore, many factors account for Grenada's social vulnerability. Evidence suggests that social vulnerability is triggered by poverty and commercial sex. This is compounded with the threat of HIV/AIDS and points to the apparent interrelationship between poverty, tourism-dependency, and the spread of HIV/AIDS.

Institutional Vulnerability

Institutional vulnerability is another major challenge which surrounds the tourism-HIV/AIDS problem. Institutional vulnerability refers to 'the greater than average risk posed by the limited capacity of domestic institutions to respond to the complexity and intensity of the pressures flowing from globalisation' (Thomas 2004:3). A major aspect of institutional vulnerability stems from external forces in the global political economy. Grenada underwent SAPs in the 1980s and from 1993–95. This resulted in the 'streamlining' or 'downsizing' of the public sector and increased unemployment. Currently the human resource and organizational capacity of the country—in the public sector, private sector and civil society generally—is 'limited and stretched.'[48] There is a 'zero-growth' policy in the public sector which has been in effect since the 1990s. In the 2000 budget speech, the minister of finance confirmed that:

> This Administration remains committed to the principle of small government and a more efficient Public Service. Consequently public sector reform remains a high priority for Government. It is the principal strategy to curb increases in recurrent expenditure. The freeze on hiring in the Public Service will remain in force and we will continue to reduce the size of the Public Service, where appropriate (Government of Grenada 1999: 19–20).

Public-sector reform has had implications for human capacity in the public service. For example, in 1988 a health planning unit was established in the MOH, but it was staffed by two officers only, who were expected to conduct health planning for the entire tri-island state. Traditionally, health services in Grenada were provided mainly through public-health facilities. However, there is a growing move towards the use of private services, in keeping with the demands of international financial institutions such as the IMF and WB, which promote privatisation as a means to increase efficiency. For instance, in 1997 and 1998 value-for-money studies were carried out in the MOH which resulted in the outsourcing of some health services. User fees were also introduced in the main hospitals. Although there are built-in provisions to protect the disadvantaged, from general observation by the author, user fees have negatively affected the ability of the poor to access health care. However, outsourcing and commercialisation initiatives are ongoing in the health sector. In 2002, the GOG launched its Public Service Management Improvement Project (PSMIP) in response to 'systems deficiencies' in the public sector. A part of this project will address commercialisation of public services. This thrust toward public-sector reform and 'smaller government' is part of a larger global neoliberal project which has implications for Grenada's ability to respond to challenges, such as the interrelationship between tourism and HIV/AIDS.

The Policy Environment

Another aspect of institutional vulnerability relates to inadequate policies. While there are initiatives to prevent the spread of HIV/AIDS and care and treatment for those afflicted and affected with the disease, currently Grenada does not have a policy that addresses the interrelationship between HIV/AIDS and tourism. This was confirmed by the minister for tourism, who indicated that there is 'no procedure to follow at this time.' The minister also pointed out that in all the national, subregional, and regional meetings that she has attended, 'HIV/AIDS has not been part of the tourism agenda.'[49] In her view, it should be. This was also pointed out by the public relations officer of the BOT, who acknowledged that the BOT is not involved in addressing the issue of HIV/AIDS. In his view, the MOH is working for the country in general, so there is not much 'rhetoric' about HIV/AIDS in the tourism sector.[50] The minister for health also confirmed that 'the issue has never come to the table in that manner.'[51]

Since Grenada is heavily dependent on tourism, it becomes problematic to include HIV/AIDS prevention messages which may appear to target tourists. As the minister for tourism explained, there must be caution, since this is a sensitive area the country will have to get into. In the minister's view, since the epidemic is quite prevalent in the world, Grenada needs to put mechanisms in place so that it cannot be spread in the society. According to the tourism minister, 'We want to protect the visitor who has the disease and protect our local people at the same time.' In her view, this must be done in 'a very polite way' so as not to destroy the tourism industry, which must continue to market Grenada's 'glamour.'[52]

The public relations officer (PRO) of the BOT also shares this view. According to the PRO:

> The BOT focuses on a certain type of tourism. We deal with up market generally. The bottom line is that we would have hoped that given the type of tourism that we promote we would not find ourselves being a victim of persons coming here for the purposes that can lead to the

further spread of this deadly virus. But in our promotional work out there, we cannot talk about it. The role of the BOT is to promote the destination as a preferred Caribbean enclave. We would encourage the MOH's efforts to educate people about the disease but in the long run I don't think you would see us becoming involved in this. The statistics are frightening and I'm scared for what can happen, but tourism promotion is our business [not health].[53]

This position by the tourism sector reflects Grenada's dependency on tourism. It also reflects the disjointed approach to the question of HIV/AIDS and tourism. The PRO's views are shared by another official in the BOT, who argues that '[i]t is not our role to educate the tourist. We have to educate *our* people. A lot of tourists come here thinking Caribbean people are easy. If we do advertisements which ask them to walk with condoms, that will reinforce in their minds that we are all about sex. And the destination offers much more than that.'[54] The former permanent secretary in the MOH argued that 'policies cannot zero in on the tourists; we have to ensure that our people develop and maintain healthy lifestyles.'[55] The minister for education foresees a similar backlash to what occurred in Barbados. She argued that tourist destinations do not have any responsibility for the tourist because 'AIDS is not an infectious disease like SARS [instead] one should deal with its *own* population.'[56] However, Benoit shares another view. She points out:

> We all have to be speaking the same language. Whichever sector we belong to—HIV prevention is our business. If you're in tourism, agriculture, or wherever, what we have to get right is that HIV affects people, it affects our development, it affects our social life. If we want to promote tourism we have to promote tourism within the context of healthy lifestyles. We have to remember that first and foremost we have to do it for us. Once we do it for us then it is available for visitors. If we can talk the same language and walk the same street I don't think we would go wrong. I don't think that because we are trying to save our country that tourists will not come to Grenada. In fact, I think that will encourage more people to come to Grenada.[57]

Therefore, one aspect of institutional vulnerability relates to conflicting goals among sectors. There appears to be a conflict between the goals of tourism marketing and promotion and HIV/AIDS prevention. For instance, there appears to be a resistance from subsectors of the tourism industry against messages that too visibly target tourists. An officer from the MOH indicated that 'there was one time we tried to set up a banner at the airport and we were told we cannot do so.'[58] As Benoit confirmed: 'I know that sometimes we attempt to put messages at the airport and we are debarred from doing so for the fear that it would affect tourism.' Benoit agreed that the MOH has to be careful with the messages that are displayed, but at the same time she argued that Grenadians have the right to protect themselves from their own selves and from visitors. In Benoit's view there is need to strike an even balance.

According to the director of the National AIDS Directorate, the tourism sector needs to find creative ways to market the tourism product that takes into account the problem of HIV/AIDS. In her view, HIV/AIDS needs to be put on 'the front burner.' The director would like to see advertisements that highlight the seriousness of the global issue. In essence, the message could be that 'we do not want you [the tourist] to be vulnerable to HIV and we do not want your presence to make our population more vulnerable to HIV. You can have the full experience, but be safe.'[59]

The Political Dimension

Another dimension of institutional vulnerability is driven by the political divisiveness within the country. Given the complexity of the problem, it is generally accepted that there needs to be a multisectoral response. In 2003, a strategic plan for HIV/AIDS was developed. One of the objectives was to facilitate a multisectoral approach to the management of HIV/AIDS. According to the minister for finance, the government considers itself an 'inclusive government' willing and ready to work with all citizens irrespective of political affiliation. The minister indicated that two mechanisms to ensure good governance include the Multipartite Consultation Committee and the Face-to-Face Programme. He indicated that the Multipartite Consultation Committee consisted of representatives from the development NGOs, the private sector, trades unions, and government. They have been engaged in developing a partnership in respect of the management of national-development issues (Government of Grenada 1999:40). The former permanent secretary, MOH described the relationship between government and the NGO community as 'cordial and collaborative.' However, Sandra Ferguson, a leading NGO representative, does not share this view. According to Ferguson:

> At this point the relationship between government and the NGO community is generally lukewarm and growing cold. There is insufficient space for civil society. Within the last five years or so international institutions have tied donor support to good governance and inclusiveness. Governments are being forced to create greater space for civil society. But in many cases this space is generally tokenism. In Grenada, we are yet to know how much the government really wants us on board. Do they want genuine participation and contributions? It has become a norm that a ministry will hardly have any major issue without sending you a document and asking for your input, or having a consultancy and inviting you to be there. That is the first stage but how this is followed up is the next step. It depends on the politicians whether they want your input or not. For example the multipartite committee which has been in existence since

1997 is going nowhere. We have seen no clear and concrete evidence of commitment to follow up or to take decisions based on our suggestions. We have to ask who sets the policy agenda? The politicians do and they decide whose views they consider and whose they don't.[60]

The minister for education notes that the extent of partisan politics in Grenada is 'extremely sad.' No country will develop with that kind of divisiveness, she argued. She went on to state that there is a current situation in Grenada in which supporters of the administration feel they can give less than their best because 'their government is in power and no one could fire them.' At the same time, opposition supporters are saying: 'We will not do anything to make this government look good.' In such a political environment, the minister argued, '[w]e are at a stand still.'[61] As the IMF noted, 'Having won less than a majority of the popular vote in the 2003 election, and with only a one-seat majority in Parliament, the government has found it difficult to build a social and political consensus on a course of action for recovery' (IMF 2005:9).

Thus, institutional vulnerability relates to limited institutional capacity, inadequate policies, sectoral conflict, and an adversarial political environment which breeds disunity. These factors converge to exacerbate the challenges posed by the interrelationship between HIV/AIDS and tourism.

Reflecting on Positive Steps

Within this environment and despite the vulnerabilities, Grenada has made efforts to address HIV/AIDS. In terms of the public-policy response, the following institutional arrangements were put in place:

- In 1984 the National AIDS Task Force was established to advise the Ministry of Health on matters concerning AIDS prevention and control and to support programme implementation.
- The National AIDS Programme was established in 1986. Its purpose was to implement the recommendations of the task force. It was designed as a unit within the Ministry of Health to provide HIV/AIDS programme management, counselling, education, condom distribution, surveillance, diagnosis, and treatment.
- A private-public sector initiative to reenergise HIV/AIDS prevention was launched in 1999. However, this initiative was not sustained.
- In 2002 the National AIDS Programme (NAP) was replaced with the National Infectious Disease Control Unit, which has responsibility for HIV/AIDS and Sexually transmitted diseases.
- GOG inputs were minimal until 2001, when EC$200,000 was budgeted explicitly for the use by the NAP.
- In terms of donor support, from 1986 to 2002 Grenada benefited from assistance from the United Nations HIV/AIDS (UNAIDS), the

Caribbean Epidemiological Centre (CAREC), the Pan American Health Organisation (PAHO), the French Technical Mission, and the British High Commission and PANCAP.

THE WAY FORWARD

Grenada is moving forward to address the problem of HIV/AIDS with assistance from the WB and other donors. In February 2003, an EC$18 million WB HIV/AIDS prevention and control project was launched in Grenada, which is being collaborated between the MOH and the prime minister's office. The WB project includes (US$130,000) for advocacy and US$128,000 for behavior change. The population groups being targeted include commercial sex workers and hotel and tourism workers. US$604,300 was earmarked for condom distribution. The project aims to promote the visible access to condoms in hotels, motels, restaurants, and other public places. The project also includes US$1,808,500 for institutional development and management and US$770,000 for treatment. In this regard, the project supports the newly established National HIV/AIDS Action Council (NAC), whose mandate is to provide policy guidance and leadership and, through its directorate, ongoing management of the program. The NAC falls under the prime minister's portfolio, since HIV/AIDS 'is considered one of the most important issues facing the country.'[62] As the prime minister further noted: 'Leadership is important, hence the decision to place the NAC in the Prime Minister's Office to give it a high level of importance.'[63] Members of the NAC are appointed by the prime minister on advice of a core group of HIV/AIDS stakeholders. The NAC reports to the prime minister and comprises representatives from:

- Persons Living with HIV/AIDS (PLWA)[64]
- Faith-based organizations
- Grenada Media Workers' Association
- Grenada Trade Union Council
- National Youth Council
- Ministry of Health (representing the government of Grenada)
- Nongovernmental organisations
- Grenada Chamber of Industry and Commerce

The National AIDS Directorate (NAD) is the executive arm of the NAC. The director relies on the National Infectious Disease Control Unit (NIDCU) within the MOH for management of all of the technical and clinical aspects of the project. The mission of the NIDCU is to

- reduce the burden of serious infectious diseases on Grenadian society;
- arrest the progress of epidemics of selected diseases which threaten the public's health;

- provide surveillance and a management system for confronting future infectious diseases which may emerge.

The HIV/AIDS programme is administered by the NAD, which has full executive responsibility for the management of the AIDS epidemic in Grenada. The NAD is a strategic unit to encourage greater synergy between the HIV/AIDS policy community and other sectors. The director of the NAD reports to the permanent secretary, prime minister's office. There is a working relationship with the Ministry of Health. NAD is responsible for the overall response. The MOH has responsibility for a coordinating and implementing role. The emphasis of NAD is on building capacity in civil society and strengthening the health centre. The work program of the NAD also includes nonhealth issues such as tourism. A significant development is the planned appointment of a tourism coordinator in the NAD. That person will be the link to pinpoint issues which relate to tourism and HIV/AIDS. According to the director of the NAD, a tourism coordinator was expected to be appointed before the end of 2006 to work with the HIV/AIDS policy community and the tourism sector to put some basic structures in place. Some of the duties of the HIV/AIDS–tourism coordinator will include:

- Working with the MOH and tourism and other stakeholders to place the issue of tourism and HIV/AIDS on the national agenda;
- HIV/AIDS training, retraining, and certification of all stakeholders in the tourism industry;
- Broad-based sensitisation programmes on the issue of tourism and HIV/AIDS;
- Increasing access to condoms in places frequented by tourists;
- Encouraging hotels and guesthouses to have condom-vending machines onsite;
- Increasing access to voluntary counselling and testing;
- Researching the linkages between tourism and HIV/AIDS, particularly issues such as sex tourism and the relationship between poverty, gender, tourism, and HIV/AIDS;
- Monitoring and evaluating projects and programs.

The minister for health indicates that in the short to medium term, there is expected to be improved collaboration. In the minister's view, Grenada has an emerging model that would lend itself to best practice. She believes that there is very strong political will, technical expertise in the NAD and the NIDCU, and a multisectoral partnership which includes the churches and other NGOs. According to the minister, '[f]or a long while Grenada was "stuttering," making two steps forward and two backward, but the country is now shaping a "collaborative model" .'[65] Benoit sees hope. Benoit is particularly pleased with the formation of the support group by PLWA and the proposed tourism coordinator in the NAD. In her view, Grenada would not be able to eradicate the epidemic, but it can be under control. If not, she fears that it could threaten the very tourism industry on which the region depends and derail the destiny of Caribbean peoples.

CONCLUSION

New perspectives in Caribbean tourism must include an analysis of the inter-relationship between tourism dependency and the threat the HIV/AIDS pandemic poses to sustainable development. This chapter attempted to do so using the case of Grenada. As the discussion showed, the interplay between tourism and HIV/AIDS presents a grave challenge for Caribbean policymakers. The Caribbean is the most tourist-dependent region in the world, and it has the second highest HIV/AIDS prevalence rates. There appears to be a direct relationship between tourism and the spread of the HIV/AIDS. This relationship is shaped by several factors: the power differentials between tourists and locals, which is historical; the 'hedonistic' nature of tourism; cultural norms; the consequences of SAPs, such as unemployment, poverty, and further weakened institutional capacity; deficiencies in governance, such as disjointed or absent policies. The convergence of these factors brings to the fore the vulnerability of Caribbean states.

Therefore, one conceptual approach to understand the problem is through Thomas's (2004) vulnerability typology, which shows the interrelationship between economic, environmental, social, and institutional vulnerability. As the case of Grenada shows, an open, dependent economy is exposed to external shocks. With limited financial and human resources it is difficult to adequately respond to such shocks. At the same time, weakened social and institutional structures undermine productivity and stifle economic growth, which further perpetuates economic, social, and institutional vulnerability. To compound the problem, small states, such as Grenada, are also exposed to environmental shocks, such as Hurricane Ivan. The severe impact of the hurricane further weakened the socioeconomic fabric and institutional framework of the country.

It can be concluded, therefore, that the vulnerability of the Caribbean is shaped by interrelated external and internal factors. Within this context the problem of tourism dependency and the threat of HIV/AIDS is complex and threatens to undermine security and development. As the region continues to search for a viable path to development, the HIV/AIDS pandemic is an unwelcome guest which inadvertently accompanies the tourism product. The challenge is to find policies which frontally address the issue without damaging the tourism industry. In the wake of the 2007 CWC, there is an urgency to tackle the tourism-HIV/AIDS dilemma. Theoretically, Grenada's emerging model is a step in the right direction. There appears to be political will; plans for greater collaboration between the HIV/AIDS prevention community and the tourism sector; the involvement of PLWAs and other stakeholders; regional and donor support. The challenge will be to convert plans into action within Grenada's political, economic, and sociocultural environment.

The way forward will be bumpy at best if the following conditions are not in place. First, the political culture needs to be transformed to ensure genuine participation and inclusion of all stakeholders. The dilemma cannot be effectively tackled within a divisive and conflictual political environment. Second, economic strategies need to be less dependent on tourism. In this context, Grenada can benefit from its membership in the CSM. There needs to a visionary and collective approach to Caribbean development which carves out creative niches which reduce vulnerabilities and maximize the economic potential of the region. Third, public education is an imperative. Ordinary people must be involved in the process. The tourism-HIV/AIDS predicament centres around the lived experiences of ordinary people. They can contribute meaningfully to finding answers 'on the ground' to address the dilemma. Fourth, the way forward also requires a cultural revolution to change behaviour. Families, churches, schools, and society as a whole must be involved in a positive way to diffuse the time bomb that is ticking. This will not happen overnight. As Grenada seeks to rebuild after Hurricane Ivan, this is an opportune time to do so correctly. There is a key strategic role for the Agency for Reconstruction and Development (ARD) in this regard.

While this is not an exhaustive study, it intends to contribute to the discourse on tourism development in the context of globalisation and deeper Caribbean integration. It can also be a useful policy tool as the Caribbean in general and Grenada in particular search for answers to the contradictions of development in this complex global era.

NOTES

1. In 2007 eight Caribbean countries hosted the Cricket World Cup 2007. The eight countries were Antigua and Barbuda, Barbados, Grenada, Guyana, Jamaica, St. Lucia, St. Christopher and Nevis, and Trinidad and Tobago.
2. Recent statistics reveal that in 2004 international tourist arrivals reached an all-time record of 763 million. Correspondingly, in absolute figures, worldwide earnings on international tourism reached a new record value of US$623 billion for that same year (UNWTO 2006). In the case of the Caribbean, over 20 million tourists visit the region each year. Between 1970 and 2000, the number of Caribbean stay-over arrivals increased five times, from roughly 3.5 million to 17.2 million (CTO 2002). International tourist arrivals in 1990 stood at 11.4 million. There was a steady increase to 18.2 million tourist arrivals in 2004 with corresponding receipts of US$19.2 billion (UNWTO, 2006).
3. In 2006 there was an estimated 39.5 million people living with HIV/AIDS in the world: an estimated 24.7 million in sub-Saharan Africa (prevalence rate 5.9 percent); 460,000 in the Middle East and North Africa (0.2 percent); 7.3 million in South and Southeast Asia (0.6 percent); 1.4 million in North America (0.8 percent); 740,000 in Western and Central Europe (0.3 percent); 1.7 million in Latin America (0.5 percent); and 250,000 in the Caribbean (1.2 percent; UNAIDS 'Global Summary of AIDS Epidemic December 2006')

(http://data.unaids.org/pub/EpiReport/2006/02-Global_Summary_2006_Epi-Update_eng.pdf). Accessed January 26, 2007, pp 1–2.

4. Peggy McEvoy, former team leader, UNAIDS Caribbean, UNAIDS. 'Special Report: AIDS in the Caribbean and Latin America' 2006.

5. In January 2006 the CSM was launched with six members: Barbados, Belize, Guyana, Jamaica, Suriname, and Trinidad and Tobago. In June of that same year, Antigua and Barbuda, Dominica, Grenada, St. Christopher and Nevis, St. Lucia, and St. Vincent and the Grenadines became members.

6. See Irene Medina, 'Eye on the CSME: CSME an Open Door for AIDS,' *The Barbados Nation* 7, November 2005.

7. 'Sex Tourism Cause HIV spread, Says T&T Minister,' *The Weekly Gleaner* (Jamaica), February 19, 2003. See also Anna Boodram, 'The Beach Bum Phenomena,' *Caribbean Voice* 3, August, 2003, and Juile Bindel, 'The Price of a Holiday Fling,' *Guardian* (London), July 5, 2003.

8. 'AIDS Activists Worried over Child Prostitution in Dominican Republic,' *Boston Haitian Reporter,* January 31, 2003.

9. amfAR is one of the world's leading nonprofit organizations dedicated to the support of HIV/AIDS research, HIV prevention, treatment education, and the advocacy of sound AIDS-related public policy. See http://www.amfar.org/cgi-bin/iowa/amfar/record.html?record=1.

10. amFAR, 'Special Report: AIDS in the Caribbean and Latin America,' p. 2 (http:www.amfar.org/cgi-bin/iowa/programs/globali/record.html?record=128). Accessed January 27, 2007.

11. The dispute began in September 1994 when Chiquita petitioned the United States trade representative (USTR) to initiate Section 301 investigations against banana policies of the EU. The United States, acting on behalf of Guatemala, Honduras, Mexico, and later Ecuador, challenged the EU regime at the WTO, based on what the United States referred to as discriminatory EU trade policy. Despite vigorous negotiation efforts on the part of the EU and ACP states, public appeals by Nongovernmental organizations and other interests groups and wide media coverage, the WTO ruled in favour of the United States in September 1997.

12. Ministry of Foreign Affairs, The Kingdom of Thailand (http://203.150.20.51/web/22.php). Accessed October 13, 2006.

13. See Sanka Price, 'Sex Tourism in for Scrutiny,' *Daily Nation* March 7, 2002.

14. 'Warning Welcome: AIDS Message to Greet Visitors,' *Sunday Sun* 21, April 2002. See also 'A Word to the Wise,' *Weekend Nation* 26, April 2002.

15. Johnson Johnrose, 'Sex, 'Secret' Finally Out,' *Daily Nation* 24, April 2002.

16. Ibid.

17. Ibid.

18. An official from the Caribbean Tourism Organisation (CTO), who requested to be anonymous, informed the author that there is no policy statement from the CTO on the question of tourism and HIV/AIDS in the Caribbean (January 2007).

19. World Bank, Grenada Data Profile. (http://devdata.worldbank.org/external/CPProfile.asp?SelectedCountry=GRD&CCODE=GRD&CNAME=Grenada&PTYPE=CP). Accessed January 27, 2007.

20. George I. Brizan, Interview by author, St. George's, Grenada, July 24, 2004.

21. See Ronald Sanders (2005), *Crumbled Small: The Commonwealth Caribbean in World Politics.* London: Hansib Publishing Ltd.; Caroline Thomas and Peter Wilkin (2004), 'Still Waiting after All These Years: "The Third World" on the Periphery of International Relations,' *British Journal of Politics and International Relations* 6: 241–58; Ulf Olsson et al., 'Special Section: Small States in the World of Markets,' *New Political Economy,* 8 (1),

2003; Commonwealth Advisory Group (1997), *A Future for Small States Overcoming Vulnerability*. London: Commonwealth Secretariat; G. K. Helleiner, 'Why Small States Worry: Neglected Issues in Current Analyses of the Benefits and Costs for Small Countries Integrating with Large Ones,' Oxford: Blackwell Publishers Ltd., 1996; R. Edis, 'Punching above Their Weight: How Small Developing States Operate in the Contemporary Diplomatic World,' *Cambridge Review of International Affairs* V (2), Autumn/Winter 1991; Clive Y. Thomas, *The Poor and the Powerless*. New York: Monthly Review Press, 1988.

22. The Eastern Caribbean Currency Area has a common Central Bank, the Eastern Caribbean Central Bank, a common currency pegged to the U.S. dollar at 2.70 per US$1 since July 1976. The ECCU comprises Antigua and Barbuda, Dominica, Grenada, St. Kitts and Nevis, St. Lucia and St. Vincent, and the Grenadines.

23. See Government of Grenada, Agency for Reconstruction and Development, 2004.

24. Eastern Caribbean dollar (EC$) is equivalent to US$2.7.

25. K. C. Mitchell, prime minister of Grenada. Interview by author, August 11, 2003, St. George's, Grenada.

26. Clarice Modeste-Curwen, former minister for health, government of Grenada. Interview by author, St. George's, Grenada, August 14, 2003.

27. Organisation of Eastern Caribbean States, 'Grenada Macro-Socio-Economic Assessment of the Damages Caused by Hurricane Ivan,' September 2004, p. 37.

28. Leroy Noel, 'Food for Sex in Grenada,' Carib Net News, September, 2004.

29. Modeste-Curwen, 2003.

30. Judy Benoit, official, Ministry of Health, Grenada, interview by author, St. George's, Grenada, September 28, 2006.

31. Claris Charles, minister for education, government of Grenada, interview by author, St. George's, Grenada, September 28, 2006.

32. Benoit, 2006.

33. Modeste-Curwen, 2003.

34. Sandra Ferguson, Executive Director, Agency for Rural Development (ART), Grenada, interview by author, August 14, 2003.

35. Cecilia Alexis-Thomas, director, National AIDS Directorate, Grenada, interview by author, St. George's, Grenada, October 2, 2006.

36. Brenda Hood, minister for tourism, government of Grenada, interview by author, St. George's, Grenada, September 28, 2006.

37. Modeste-Curwen, 2003.

38. Alexis-Thomas, 2006.

39. Cedriann J. Martin, 'Caribbean Sex Market and Economy,' *Trinidad and Tobago Express*, September 23, 2006.

40. Ann Antoine, minister for health, government of Grenada, interview by author, St. George's, Grenada, September 28, 2006.

41. Grenada National Strategic Plan for Health. Situation Analysis, 2005.

42. Charles, 2006.

43. Mitchell, 2003.

44. Joseph Charter, former permanent secretary, Ministry of Health, interview by author, St. George's, Grenada, August 13, 2003.

45. Modeste-Curwen, 2003.

46. Antoine, 2006.

47. Charles, 2006.

48. Government of Grenada, 'Public Sector Modernisation 2006–2010,' draft, November 2005.

49. Hood, 2006.

50. Edwin Frank, public relations officer, Grenada Board of Tourism, interview by author, St. George's, Grenada, October 2, 2006.
51. Antoine, 2006.
52. Hood, 2006.
53. Frank, 2006.
54. Interview by author with an official of the Grenada Board of Tourism who requested to be anonymous
55. Charter, 2003.
56. Charles, 2006.
57. Benoit, 2006.
58. Officer of the Ministry of Health, interview conducted by author. Officer requested anonymity.
59. Alexis-Thomas, 2006.
60. Ferguson, 2003.
61. Charles, 2006.
62. Mitchell, 2003
63. Ibid.
64. People Living with AIDS (PLWA) is an active group which celebrated its fifth anniversary in 2005. The group is funded and benefits from treatment.
65. Antoine, 2006.

REFERENCES

Alexis-Thomas, C. Director, National AIDS Directorate, Grenada. Interview by author, St. George's, Grenada, October 2, 2006.
Allen, C., R. McLean, and K. Nurse. 2004. The Caribbean, HIV/AIDS and security. In *Caribbean security in the age of terror,* I. Griffith, ed. Jamaica: Ian Randle Publishers.
Antoine, A. Minister for health, government of Grenada. Interview by author, St. George's, Grenada, September 28, 2006.
Barrow, V. M. 2003. The politics of tourism development in the Caribbean: Insights from the Barbados case. Unpublished master's thesis, University of the West Indies, Cave Hill.
Beckles, H. 1989. *Natural rebels: A social history of enslaved black women in Barbados.* London: Zed Books.
Benoit, J. Official, ministry of health, Grenada, interview by author, St. George's, Grenada, September 28, 2006.
Borland, R. et al. 2004. International Organisation for Migration, HIV/AIDS and Mobile Populations in the Caribbean. A Baseline Assessment. Santo Domingo, Dominican Republic.
Boxill et al. I. 2005. *Tourism and HIV/AIDS in Jamaica and the Bahamas.* Kingston: Jamaica: Arawak Publications.
Brennan, D. E. 1989. Everything is for sale here: Sex tourism in the Soscia, the Dominican Republic. Unpublished doctoral dissertation, Yale University, New Haven, CT.
Brizan, G. I. 1984. *Grenada island of conflict from Amerindians to people's revolution 1948–1974.* London: Zed Books.
———. 2004. Interview by author, St. George's, Grenada, July 24.
Camara, B. 2001. 20 years of the HIV/AIDS epidemic in the Caribbean: A summary. Port-of-Spain, Trinidad: CAREC/PAHO/WHO.
Caribbean Community, the Caribbean Regional Strategic Framework for HIV/AIDS 2002–2006. 2000. Georgetown, Guyana: CARICOM Secretariat.

Charles, C. Minister for education, government of Grenada, Interview by author, St. George's, Grenada, September 28, 2006.

Charter, J. Former permanent secretary, Ministry of Health, Interview by author, St. George's, Grenada, August 13, 2003.

Ferguson, S. Executive director, Agency for Rural Development, Grenada, interview by author, August 14, 2003.

Frank, E. Public relations officer, Grenada Board of Tourism, Interview by author, St. George's, Grenada, October 2, 2006.

Government of Grenada. 2000. Budget Speech. St. George's, Grenada, 10 December 10, 1999.

———. 2001 Budget Speech, St. George's, Grenada, January 12, 2001.

———. 2002. HIV/AIDS Prevention and Care Project Project Operations Manual.

Government of Grenada. '2006 Budget Speech,' St. George's, Grenada December 15, 2006.

———. 2003. Strategic Plan for HIV/AIDS in Grenada.

———. 2004. Agency for Reconstruction and Development, Report on the Reconstruction and Development Programme.

———. 2005 Budget Speech., St. George's, Grenada, April 11, 2005.

———. . 2005. National Strategic Plan for Health (2006–2010). Concept Paper— Health for Economic Growth and Human Development..

Grenada Board of Tourism. 2004. Annual Statistical Report 2004: St. George's, Grenada.

Grenada. 2005. Article IV Consultation—Staff Report; and Public Information Notice on the Executive Board Discussion. August. In IMF Country Report No. 05/290: International Monetary Fund

Hood, B. 2006. Minister for Tourism, Government of Grenada, interview by author, St. George's, Grenada, September 28.

International Monetary Fund. 2005. Grenada: 2005 Article IV Consultation-Staff Report; and Public Information Notice on the Executive Board Discussion. IMF Country Report No. 05/290 August 2005.

Karch, C. A., and G. Dann. 1981. Close encounters of the third kind. *Human Relations* 34:249–68.

Kempadoo, K. 1999. *Sun, sex and gold: Tourism and sex work in the Caribbean.* New York: Rowman and Littlefield.

Klak, T. 1998. *Globalization and neoliberalism: The Caribbean context.* New York:: Rowman and Littlefield.

Mitchell, K. C. 2003. Prime minister of Grenada, interview by author, St. George's, Grenada, August 1.

Modeste-Curwen, C. 2003. Former minister for health, government of Grenada, interview by author, St. George's, Grenada, August 14.

Morrissey, M. 1989. *Slave women in the New World: Gender stratification in the Caribbean.* Lawrence: University Press of Kansas.

Naisbitt, J. 1994. *Global paradoxes: The bigger the world economy, the more powerful its smallest players.* London: Bealey.

O'Connell Davidson, J. 1996. Sex tourism in Cuba. *Race and Class* 38, 39–48.

Ohmae, K. 1990. *The borderless world: Power and strategy in the interlinked economy.* London:: Fontana.

Organisation of Eastern Caribbean States. Grenada: Macro-Socio-Economic Assessment of the Damages Caused By Hurricane Ivan September 7, 2004. St. Lucia: Organisation of Eastern Caribbean States.

Organisation of Eastern Caribbean States Toward a New Agenda for Growth. Report No. 31863-LAC April 7, 2005. World Bank.

Organisation, United Nations/World Tourism. 2006. Tourism Highlights 2005 Edition (http:www//unwto.org/facts/menu.html). Accessed December 2006.

Sullivan M. P. 20 June 2006. HIV/AIDS in the Caribbean and Central America: CRS Report for Congress.

Poverty Assessment Report, Grenada, Vol 1. Main Report Tunapuna. 1999. Trinidad and Tobago: Caribbean Development Bank, Kairi Consultants Ltd. and National Assessment Team of Grenada

Pritchard, A., and N. J. Morgan. 2000. Privileging the male gaze: Gendered tourism landscape. *Annals of Tourism Research* 27 (4):891–910.

Thomas, C. Y. 14 January, 2004. Small states and the vulnerability debate, Guyana and the wider world. *Stabroek News*.

UNAIDS Special Report: AIDS in the Caribbean and Latin America. 2006. UNAIDS.

UNAIDS/World Health Organisation Report on the global AIDS epidemic. 2006. UNAIDS/WHO.

Wood, N., ed. 2000. *The political economy of globalization.* New York: St. Martin's Press.

World Bank. 2005. Organisation of Eastern Caribbean States, Toward a New Agenda for Growth. Report No. 3/863-LAC. Washington, DC, USA.

11 Regional Partnerships
The Foundation for Sustainable Tourism Development in the Caribbean

Leslie-Ann Jordan

INTRODUCTION

In light of the many challenges facing small island developing states (SIDS) in the Caribbean in the twenty-first century, many of them have been motivated, or in some cases compelled, to identify policy alternatives which are not only effective and efficient but which also provide the greatest potential to affect the desired economic and social outcomes. Consequently, they have identified tourism as an engine of growth. However, the worldwide tourism industry is becoming increasingly more competitive, not only because travellers are becoming more discerning in choosing their destinations but also because the number of destinations to choose from is multiplying as more governments identify tourism as an economic tool for development.

Given the global reality, many SIDS are discovering that partnerships are not only a necessary requirement, but they are vital to achieving their sustainable tourism development goals. Much of the existing literature on partnerships has examined microlevel issues such as partnerships between private and public sector, within private sector and public sector, or partnerships with stakeholders such as airlines, hotels, and other stakeholders in the corporate marketplace (Jamal and Getz 1995; Reed 2000) . However, previous research on tourism in the Caribbean has neglected to examine the importance of regional partnerships and their role in achieving sustainable tourism development. Consequently, this chapter addresses this existing gap from a macroperspective by examining tourism partnerships in light of two Caribbean realities: economic realities such as the Caribbean Single Market and Economy (CSME) and other such trading agreements, and historical realities of dependency, insularity, fragmentation, and continued calls by some small island states for autonomy.

The chapter also discusses two pertinent questions. First, how can Caribbean islands address the need for regional integration and partnership within the tourism industry while, at the same time, differentiating themselves from their neighbours by celebrating and advertising their uniqueness? And second, what is the role and function of regional tourism organizations such as the Caribbean Tourism Organization (CTO) and the Caribbean Hotel Association (CHA) in facilitating such partnerships?

STRENGTH OF REGIONAL TOURISM PARTNERSHIPS

Regional tourism partnerships can be a powerful tool for promoting sustainable tourism development in the Caribbean. According to the United Nations (2006), partnerships are voluntary multistakeholders initiatives which contribute to the implementation of intergovernmental commitments. Many researchers have documented the benefits of partnership arrangements in sustainable tourism development (Araujo and Bramwell 2002; Jordan 2004; Timothy 2004; United Nations 2006). Partnerships have the potential benefits of shared information and decision making; the development of appropriate tourism policies that reduce the negative impacts of tourism; efficient administration of tourism; decrease in misunderstandings and conflicts between stakeholders related to overlapping and conflicting roles and responsibilities; and efficient use of scarce resources (Bramwell and Lane 1999; Araujo and Bramwell 2002; Jordan 2004). Enabling all relevant stakeholders to make concrete contributions to sustainable tourism development can improve the timeliness and effectiveness of policies and projects.

The rationale for establishing regional tourism partnerships in the Caribbean is clear:

- To ensure a more effective and efficient use of regional resources, channelling available resources into the region's key tourism priorities;
- To develop a regional tourism delivery structure that is fully inclusive of the whole region;
- To provide a more coordinated and cohesive approach to tourism development throughout the region;
- To address sustainability issues as international funding declines;
- To gather and disseminate market intelligence, business opportunities, policies, and strategies and best practices.

To a large extent, the success of sustainable tourism development in the Caribbean depends on regional partnerships, which will bring together a wide range of stakeholders representing interests at the regional, national, and local levels. Acquiring knowledge and appreciation of the complexity of regional partnerships and its environment would help to:

- facilitate better interorganisational relationships between tourism stakeholders by uncovering historical and cultural sensitivities;
- expose differing organisational values;

- highlight areas of overlap and ambiguity in terms of roles and responsibilities of organisations and actors; and
- provide the relevant background and context within which constructive and productive dialog can occur in order to address and resolve some of the tourism priorities of the region.

In the Caribbean, cooperation and partnership are needed to give shape to the regional tourism industry while still maintaining the strength of a diversified tourism product. Given the global climate, partnerships are being used more and more in tourism to achieve business and community goals and to ensure a high level of competitiveness and a high degree of sustainable tourism development.

Beyond simple networks, partnerships require the commitment of the participants to work fully together to address problems and opportunities. This means they must accept long-term structures that work towards sustained commitment to change and the achievement of quality. They must also accept an active commitment to changing the internal operations of each participant and helping other participants to change to achieve an improved system overall (Mackintosh 1992).

However, research has shown that there are significant challenges with partnership approaches to sustainable tourism development (Bramwell and Lane 2000; Hall 2000; Araujo and Bramwell 2002). Issues of power, trust, limited time, and financial resources and stakeholder conflicts are major hindrances to collaboration. Although these are real concerns, the forces of globalization and economic restructuring dictate that the region seriously considers partnerships as a vehicle for promoting sustainable tourism.

ECONOMIC REALITY: GLOBALISATION AND ECONOMIC RESTRUCTURING—IMPACT ON THE CARIBBEAN

The increased dependence in the region on tourism is due in large part to the forces of globalisation, which have marginalized many of the region's traditional export sectors such as banana, copra, cocoa, coffee, and sugar. Improved communications and transportation technologies have facilitated the movement of goods, people, and ideas across great distances and between widely diverse cultures; and, as a result, national barriers to the movement of goods, services, investment capital, and money are being eliminated, and government intervention in economic life is being kept to a minimum (Pantin 1995; Girvan 1999; Timothy 2004). This reality has drastically affected the way that people live and do business nowadays.

In this global economy, Caribbean SIDS are being forced to liberalise their economies at the fastest rate possible in order to facilitate foreign investment. For example, the North American Free Trade Agreement

(NAFTA) requires prospective participants to liberalise their entire economies as a precondition to even being invited for membership. In terms of
regional partnerships, Girvan (1999) succinctly highlighted the challenge
posed to Caribbean countries attempting to respond to the forces of globalisation, when he declared that

> First, globalisation *requires,* as a necessary survival and development
> strategy, greater co-operation and integration within the Caribbean re-
> gion. Second globalisation *generates,* as a result of its dynamics and
> content, tendencies towards renewed differentiation and fragmentation
> within the Caribbean. Resolving this contradiction is an historic chal-
> lenge for the region (Girvan 1999: n.p.).

Girvan's statement illuminates one of the recurring dilemmas faced by
Caribbean islands that recognise the need for regional integration and cooperation but at the same time also feel the need to differentiate themselves
from their neighbours by celebrating and advertising their uniqueness as
part of their competitive advantage. As the St. Kitts–Nevis Constitutional
Commission astutely observed, "There is a strange political irony roaming
the world these days. On the one hand, globalisation is everywhere pro-
claimed, and a stately march continues towards integration . . . On the other
hand, and simultaneously, the insistence on separation has not abated . . .
(SKN Constitutional Commission 1998:1). This phenomenon could explain
why it has taken the region so long to agree on using a Caribbean brand for
tourism marketing purposes.

With globalisation and the new world economic order has come a repo-
sitioning of the Caribbean region on the international agenda (Pantin 1995;
Jessop 1999; Payne and Sutton 2001). The question now being posed in
the new millennium is: Is anyone in Europe or North America really interested in what happens to the Caribbean? (Jessop 1999; Morrison 2006).
The very fact that the question is being posed indicates that there has been
a significant shift in global thinking and operations. The end of the Cold
War has brought an end to regular cooperation between the Caribbean ad
the United States and has significantly altered the character and the scope of
U.S.-Caribbean relations (Griffith 2000; Meighoo 2000; Payne and Sutton
2001; Morrison 2006). It is now difficult to identify any other issue that
might create a positive policy towards the Caribbean (Richardson 1992;
Jessop 1999).

As a result of these events, many policymakers are now realising that
the answer lies within the Caribbean itself. No one will take the region
seriously or work with it in the future unless it is unified, has coherent
policies, and is engaged in a coordinated way in putting forward clear
ideas that would ensure sustainable economic growth and development
(Jessop 1999; Holder 2001; Payne and Sutton 2001). In this respect,
one of the major challenges facing the region is that it still relies on and

utilises old models to try to deliver new solutions in a totally changed global environment.

Jessop (1999: n.p.) suggested that "no one outside of the Caribbean is interested any longer in playing the role of the seventh cavalry riding to the rescue of the region. The Caribbean has entered the true age of independence." In the same vein, former Barbadian Prime Minister Owen Arthur asserted that "It must by now be clear that we [the Caribbean] can no longer cling to any misguided notions that our economic fortune will be any longer buttressed by the dubious largesse of former colonial powers" (Gibbings 2000: n.p.). Jamaica's then foreign minister, Seymour Mullings, made the point more forcefully:

> We are witnessing, at this point in our history, the relentless march of the forces of globalisation presenting us with unfamiliar challenges which threaten to overwhelm us individually . . . We in the Caribbean have long recognised that our survival as viable individual entities can only be assured if we capitalise on the full potential of our collective capacities (Gibbings 2000: n.p.)

The forces of globalisation have also been joined by those of economic restructuring, which operate together to place immense pressure on SIDS. The formation of economic blocs around the world is one significant manifestation of this. The Caribbean found itself increasingly marginalised as the world rearranged itself into these free-market trading blocs. For example, the European Union (EU) now boasts a unified market of 320 million consumers; the Association of Southeast Asian Nations (ASEAN), with about 420 million people, has agreed to create a free-trade area as a precursor to the establishment of a common market; and NAFTA has an annual production of over $6 trillion and some 37 million consumers in Canada, Mexico, and the United States (Griffith 2000; Timothy 2004).

One of the consequences of this megabloc phenomenon for the Caribbean is the potential reduction or even loss of economic assistance, foreign investment, and preferential trading deals (Pantin 1995; Griffith 2000; Payne and Sutton 2001). Trading blocs such as the NAFTA and the EU have made it increasingly difficult for traditional Caribbean manufacturing and agricultural industries to remain competitive, giving tourism an even higher priority throughout the region. Such conditions both regionally and internationally encourage tourism expansion more than ever before.

Given this economic reality, regional partnerships can be used as a pathway to sustainable tourism development. Unfortunately, the region's historical legacy of fragmentation, suspicion, insularity, and dependency has continued to be a stumbling block to regional integration. The reality is that any dialogue about regional tourism partnerships must be conducted in the context of the region's history.

HISTORICAL LEGACY: DEPENDENCY, INSULARITY, AND FRAGMENTATION

The Caribbean region is fascinating because of the many lessons that can be drawn from its history. Intercolonial rivalries carved up the region and led to its present-day fragmentation in language and politics (Richardson 1992; Grugel 1995). West Indian history tells a tale of how rivalries among West Indian islands have long inhibited cooperation and thwarted wider national or supranational aims (Watson 1979; Lowenthal and Clarke 1980). The Caribbean islands' desire to be on their own has been influenced by geographical isolation and colonial dependence. Williams (1970:129) explained that it was the constant conflict of interests between colonies and metropolitan country and between colony and colony that bred the "inordinate selfishness and inveterate particularism of the British West Indian planters" and indeed the island residents. Consequently:

> French, Dutch, and British linguistic and administrative barriers made St. Vincent almost unknown to Martinique, and Anguilla just a myth in Aruba; Caribbean appendages of the same European empire usually knew more of the mother country than of their West Indian neighbours (Lowenthal and Clarke 1980:294).

In the late nineteenth century, small islands in the Caribbean had not yet developed close contacts amongst themselves. Communications between islands were poor and interisland trade was very limited. As such, West Indians lacked a strong sense of regional cohesion and thought mainly in terms of their individual islands and the mother country (Watson 1979). This West Indian parochialism was based on entrenched suspicion and fear of neighbours as competitors, rivals, or agents of sedition (Lowenthal and Clarke 1980). Even well into the twentieth century, almost every island in the British West Indies remained deeply suspicious of enforced links with others and was slow to appreciate the advantages of cooperation and joint programmes for development (Watson 1979).

Consequently, at the international level, the question has been and remains: Why is the Caribbean so divided and above all why has there not been better integration? This is a pertinent question given the fact that nowadays, cross-border sustainable tourism planning is increasing in many parts of the world as neighbouring destinations realize the value of working together to develop and promote common resources. At the regional and international level, strategic allegiances, such as the EU, NAFTA, and CARICOM (Caribbean Community), are examples of multilateral cross-border cooperation that are becoming more commonplace (Timothy 1998:55). While the author is in no position to prescribe exactly what decisions or actions need to be taken, there is support for calls for the Caribbean to "reinvent" the manner in which it approaches sustainable tourism planning and policymaking. This chapter is not suggesting political reform or unification, but rather it is advocating for pragmatic solutions for sustainable tourism policy which are reasonable and appropriate. Caribbean countries need to re-examine the structure of their tourism ministries and/or authorities, as well as their intraregional partnerships, in order to make them more relevant to today's dynamic environment.

Decision makers in SIDS can play their cards more effectively, but one of the reasons that they fail to do so is because of inertia, which Selwyn (1975:19) described as "the continued use of systems, policies, attitudes, habits of work [and institutional arrangements] for years after they have become obsolete and irrelevant." This is not a problem that only small island states are susceptible to, but in their case, its consequences may be more serious and far reaching. This chapter presents a model for regional partnerships that seeks to break this state of inertia in the region. Individual destinations and organizations can no longer afford to myopically guard their territory at the expense of regional competitiveness and survival.

COLLABORATION, CONFLICT RESOLUTION, AND COORDINATED EFFORT: CORNERSTONES OF REGIONAL TOURISM PARTNERSHIPS

Figure 11.1 outlines a conceptual model for regional tourism partnerships that can be used as a foundation for analysis and discussion of such relationships between island nations of the Caribbean. As the model shows, it is only when regional organizations effectively collaborate, address conflict, and achieve coordinated effort will the region be able to deal with the tourism priorities and achieve regional benefits. Developing an effective strategy for sustainable tourism development in the Caribbean requires the involvement of regional trade/political organizations, regional tourism organizations, as well as national tourism organizations and other interested stakeholders. The possibility of organizing a Caribbean Tourism Council that comprises the heads of each of these groups and/or organizations is an option that the region should seriously consider, in order to address the region's tourism priorities. In order for such an arrangement to work, all stakeholders participating in this endeavour must first go through a three-step process: (1) collaboration, (2) conflict resolution, and (3) coordinated effort.

Jamal and Getz (1995) define collaboration as a process of joint decision making among autonomous, key stakeholders of an interorganisational domain to manage issues related to the planning and development of the domain. The collaborative approach is relevant to this discourse on sustainable tourism development because it can specifically help destinations where fragmented and independent planning decisions by different tourism stakeholders give rise

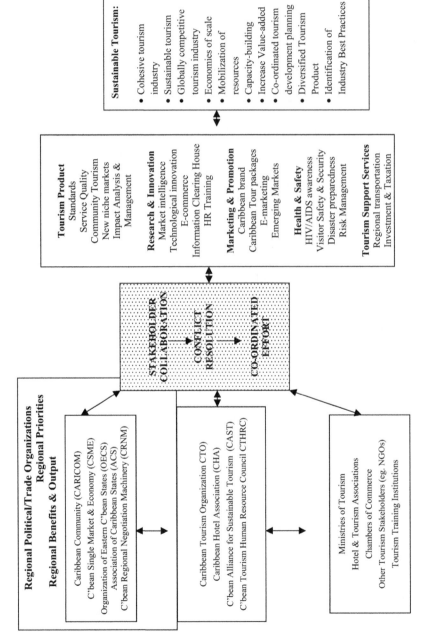

Figure 11.1　Conceptual model for regional tourism partnership development.

to power struggles over resources (Jamal and Getz 1999; Ladkin and Bertramini 2002). Gray (1989) identifies five critical features that determine the outcome of such collaborations:

1. Organisations must be seen and respected as autonomous but interdependent;
2. Solutions to problems emerge by dealing constructively with differences;
3. There must be joint ownership of decisions made during the process;
4. Organisations assume collective responsibility for the future direction of the region;
5. Collaboration is an emergent process.

There is a need for SIDS of the Caribbean to adopt these core principles when forming regional tourism partnerships. For archipelagic regions, such as the Caribbean, the collaborative planning approach is necessary, as it "attempts to act as a tool to solve the many problems that arise when there is a lack of understanding and few shared common goals between the many parties often involved in tourism" (Ladkin and Bertramini 2002:71). However, much of the discussion on cooperation and collaboration in the tourism literature often neglects the broader political context within which inter-organizational collaboration is occurring (see Bramwell and Lane 1999). This is a grave oversight, and research has shown that pleas for better communication and greater cooperation in SIDS are, more often than not, a mask of more fundamental issues, such as conflict of value systems between organisations; ambiguity of roles and responsibilities between individuals and institutions; and the need for legislative reform (Jordan 2004).

Understandably, the need for conflict resolution is evident whenever multiple groups such as public sector, private sector, local communities, and nongovernmental organisations come together. Naturally, these groups will have different interests, and, given the history and politics of the region, conflicts may arise concerning: the distribution of power; resource flows; mistrust and suspicion amongst stakeholders; social, economic, and political value systems; personality and the legislative framework. According to Hall (2000:183), conflict resolution is "a process of value changes that attempts to manage disputes through negotiation, argument and persuasion by which conflict is eliminated or at least minimized to the extent that a satisfactory degree of progress is made by the interested stakeholders." It is important that all stakeholders agree on a modus operandi that would help to mediate this conflict by providing a set of ground rules and procedures that dictate agenda setting, how and where demands can be made, who has the authority to take certain decisions and actions, and how decisions and policies are implemented (Brooks 1989; Gray 1996; Araujo and Bramwell 2002).

After conflict resolution, there must be a coordinated effort to address regional priorities. Coordination usually refers to "the problem of relating units or decisions so that they fit in with one another, are not at cross-purposes, and operate in ways that are reasonably consistent and coherent" (Spann 1979:411 in Hall 1994:33). Work coordination is also necessary to achieve interorganisational equilibrium. According to Benson (1975), work coordination involves the pattern of collaboration and cooperation between organisations whereby work is coordinated to ensure that the programs and activities between the organisations are geared into each other for maximum effectiveness and efficiency. Therefore, if the organisations outlined in Figure 11.1 are not communicating and collaborating consistently with each other, then their work coordination would be low. This would make it difficult for stakeholders to grasp a holistic vision for sustainable tourism development for the region because they would have no idea what they should be working together to achieve. This lack of direction often leads to organizations and individuals taking their own paths to development, which may not be in the best interests of the region's overall tourism development. It is therefore not surprising that common solutions, actions, and budgets have not been developed to address critical problems facing the region's tourism effort.

As Figure 11.1 shows, there are numerous regional organisations involved either directly or indirectly in the sustainable development of tourism in the Caribbean. The next section will briefly discuss the role that each of these organisations plays in developing tourism plans and policies for the region, as well as their efforts to collaboratively search for solutions for the region's largest industry.

Role of Regional Tourism Organisations

Due to space constraints, this chapter will focus in greater detail on the two regional organisations with a direct mandate to oversee tourism development in the region: the Caribbean Tourism Organisation (CTO) and the Caribbean Hotel Association (CHA). However, it is necessary to first briefly discuss the integral role of CARICOM, CSME, OECS, or the ACS as they continue their efforts to develop initiatives "... intended to benefit the people of the Region by providing more and better opportunities to produce and sell our goods and services and to attract investment" (CARICOM 2006: n.p.). These organisations have recognised that the tourism industry's importance to the region could not be overemphasised, as it contributes about 25 percent of the region's gross domestic product (GDP) and a significant share of the employment.

For example, when the CARICOM heads of government met in the Bahamas in 2001 for a two-day tourism summit, they discussed several major issues affecting the region's tourism industry, including: tourism promotion, market development, air transportation, and the need to reengineer the Caribbean tourism product specifically as it relates to price, quality, and diversity of the product and service levels (GOSKN 2001). They also agreed to launch a major advertising and marketing campaign to help counter the downturn in the industry caused by the events of September 11. The meeting included regional and international tourism officials and representatives of the CTO and the CHA.

The OECS also acknowledges that tourism is a major contributor to the overall economic activity, foreign exchange earnings, and employment of residents within the OECS. Consequently, it has adopted a three-pronged approach to tourism, including: development of joint promotion and marketing of the OECS region; development of a programme to encourage and facilitate intra-Caribbean travel; and development of a common short-term response to the demands of international airlines and the hotels, which, while being sensitive to the current situation, will not compromise the need for government to maintain revenue flows (OECS 2006).

In terms of the ACS, one of its main objectives is promoting the sustainable development of the greater Caribbean. Its current focal areas are trade, transport, sustainable tourism, and natural disasters (ACS 2006). The ACS has established a Special Committee on Sustainable Tourism, which works to ensure that destinations can attract visitors but, at the same time, do so in a way that would not harm the physical environment or the communities that surround them. It also seeks to establish criteria for sustainable tourism destinations.

The CTO, established in 1989, is the official body for promoting and developing tourism throughout the Caribbean. It is seen as an agent of change and direction and a coordinator of regional tourism activity. The CTO works to encourage sustainable tourism that is sensitive to the economic, social, and cultural interests of the Caribbean people, preserving their natural environment and providing the highest quality of service to visitors (CTO 2006). Members of the CTO include 32 member countries, including English-, French-, Spanish-, and Dutch-speaking nation-states and territories, and private companies such as airlines, hotels, cruise operators, and travel agencies. The CTO's program focuses on strategic areas such as tourism marketing, travel trade exhibitions, hosting Caribbean tourism conferences, public relations and promotions, research and information management, human resource development, product development and technical assistance, finance and resource management, and consultancy services.

On the other hand, the CHA, which was founded in 1962, actively promotes the continuing improvement and expansion of the Caribbean hospitality industry, encouraging the exchange of information through timely publications, conferences, special interest programs, training activities, and regional marketing initiatives (CHA 2006). It embraces 36 national hotel associations and their member hotels, which comprise over 1,100 member properties and 125,000 rooms. The allied membership of over 520 companies come from the industry at large, including airlines, hotel representatives, travel wholesalers, and retailers, trade, and consumer press, advertising and public relations agencies, hotels, and restaurant suppliers.

Although the CTO and the CHA have been working collaboratively in the past, it is only in 2005 that they signed a historic 14-point Memorandum of Understanding and Joint Cooperation in which the two organizations pledged "full cooperation and collaboration between the public and private sectors of the Caribbean tourism industry in order to proceed in a coherent, effective and efficient manner." According to Vincent Vanderpool Wallace, secretary general of the CTO, "we have put together this Memorandum of Understanding so that we move forward as one Caribbean because it makes us stronger together" (CTO 2005:1). Areas of cooperation identified in the MOU include tourism marketing and promotion; the preparation of a single Caribbean Web site for consumers and membership; joint research on aviation issues, cruise tourism, hotel operating study; joint efforts in human resource development and training; and a common approach to international and bilateral donor agencies like the Organisation of American States (OAS), the European Union (EU), and the Caribbean Development Bank (CDB) (CTO 2005).

The agreement should also provide a platform for activating linkages with other economic sectors, as well as increase the awareness of the Caribbean through the use of a new Caribbean logo. In late 2006, so far, six CHA charter members, four destinations, and two hotel brands were using the Caribbean logo in all their collateral material, Web sites, and advertisements. In addition, the CHA has signed a cooperation agreement with the OAS for the purpose of promoting social and economic development in the Caribbean through programmes, projects, and activities focused on helping to create and to advance efforts to improve the quality of tourism in the Caribbean. The CHA has a project office in Barbados for the management and execution of these various funded initiatives.

Given the many regional priorities (see Figure 11.1) and the challenges facing the region's tourism industry, it is imperative that these regional organisations work together to strengthen the institutional framework for tourism. A concerted joint effort would provide significant net benefits to the tourism industry, as research and statistics, resources, projects, expertise, and finances are shared. Possible overlap between CARICOM and the ACS should be monitored to reduce unnecessary competition and to ensure that key tourism policy areas do not fall through the gaps where jurisdiction is unclear (Bonnick 2000; Girvan 2000). Additionally, the CTO and the CHA need to lobby more aggressively in order to place tourism development high on the agenda of the necessary supranational governance bodies like CARICOM and OECS. For too long, tourism issues have been discussed as a side issue to other regional concerns such as transportation, capital flow and economic, environmental, and social sustainability.

All the organisations outlined in Figure 11.1 must work together to ensure that the desire for sustainable tourism development in the region receives the highest political support and cooperation from national, regional, and international organisations. This kind of support is needed in order to draft and enforce tough legislation that would protect the environment, develop consensus on the actions that need to be taken, and ensure that the main stakeholders hold equitable reserves of resources, knowledge, and power.

CONCLUSION

Undoubtedly, the Caribbean, like many developing regions around the world, is faced with many external as well as internal challenges to future tourism development. This chapter examined the issues related to the impact of globalisation and liberalisation on Caribbean economies and the challenges of operating in a competitive regional and international tourism market. In its quest to develop a sustainable tourism industry, it would be remiss of the Caribbean region if it did not examine its policies and strategies in the context of the international environment of globalisation and economic restructuring, with its calls for greater cooperation and integration in an open-market setting. This issue is relevant and timely because the Caribbean region has been gearing itself up for a world of trade liberalisation, megatrading blocs, and a dramatic increase in the trade of services such as tourism.

Confronting these challenges requires the Caribbean region to develop a strategy with two focuses. The first focus involves strengthening the bargaining position and negotiating capacity of the region's states in their external economic relations. Such a strategy may include an economic union, such as what is being proposed by the CSME, or even some sort of political confederation of Caribbean states stretching from Cuba and the Bahamas in the north to Suriname in the south (Girvan 1999). This is very much outward looking and is concerned with the region's relationship with outside powers. The organisations that are part of the regional institutional framework for tourism should be actively involved in trying to accomplish this goal.

As this chapter has discussed, the second focus is very much introspective and will involve a closer examination and assessment of intraregional relationships. The internal core-periphery framework (Jordan 2004) will be a powerful tool that can be used to conduct such evaluations. Within this dialogue, conflict, dissatisfaction, and injustice can be identified, understood, and adequately addressed. The region can no longer afford to bury its collective head in the sand. Such denial will not help to facilitate the level of cooperation that is needed to combat the forces of globalisation.

This chapter offers a broad, conceptual model for regional tourism partnership development and acknowledges that further research is required to explore the issues raised and to gather more information from relevant stakeholders. Furthermore, as institutional and partnership arrangements are constantly changing and evolving, there is a need to conduct longitudinal research, recognising that stakeholders' views change over time and that the internal and external environments that govern the development of these partnerships are dynamic.

Further research is also needed to determine what factors influence the different stages in the development of regional tourism partnerships, including problem setting, agenda setting, policy formulation, decision making, implementation, evaluation, monitoring, and maintenance. This type of research would help to facilitate a more in-depth understanding of the development and effectiveness of regional tourism partnerships by breaking down the complexity of the process into a limited number of stages. Each of these stages can then be investigated alone or in terms of its relationship to any or all of the other stages of the cycle. This will aid theory building by allowing numerous case studies and comparative studies of different stages to be undertaken.

In addition, a wider examination of the interorganisational relationships of the tourism organisations under examination would enhance understanding on institutional arrangements by focusing on any one or more of the following questions: How do these organisations resolve conflict, which has been engendered by the internal core-periphery relationship? To what extent has the internal core-periphery relationship influenced actors' values, ideologies, and priorities? What tools or techniques have been implemented to try to overcome the effects of having a weak institutional framework for tourism? How successful have they been? How successful have existing partnerships been in achieving sustainable tourism? What have been some of the challenges encountered in forming effective partnerships?

Finally, in its design of regional partnerships, the region must ask itself:

- How best could small island states in the Caribbean address problems with regional partnerships in order to respond to the new challenges they are facing?
- Is this the appropriate time for small islands like Barbuda, Nevis, and, to a lesser extent, Tobago, to be contemplating secession, or should they instead be concentrating all their efforts and resources into revising their institutional arrangements so that they become more competitive in the global marketplace?
- Are calls for self-determination more potent than the calls for national and regional integration and survival?

Clearly, most of these questions transverse the tourism dialogue and are fundamental geopolitical issues that the region will continue to grapple with well into the twenty-first century.

REFERENCES

Araujo, L. , and B. Bramwell. 2002. Partnership and regional tourism in Brazil. *Annals of Tourism Research* 29 (4):1138–1164.

Association of Caribbean States (ACS). *About the association of Caribbean states.* 2006. Online: www.acs-aec.org.

Benson, J. K. 1975. The interorganizational network as a political economy. *Administrative Science Quarterly* 20 (2):229–49.

Bonnick, G. 2000. *Toward a Caribbean vision 2020: A regional perspective on development challenges, opportunities and strategies for the next two decades.* Washington: Caribbean Group for Cooperation in Economic Development (CGCED).

Bramwell, B., and B. Lane. 1999. Collaboration and partnerships for sustainable tourism. *Journal of Sustainable Tourism* 7 (3 & 4):179–81.

———. 2000. Collaboration and partnerships in tourism planning. In *Tourism collaboration and partnerships: Politics, practice and sustainability.* B. Bramwell and B. Lane, eds. Clevedon, UK: Channel View.

Brooks, S. 1989. *Public policy in Canada: An introduction.* Toronto: McClelland & Stewart Inc.

Caribbean Community (CARICOM). *Single market.* 2006. Online: www.caricom.org.

Caribbean Hotel Association (CHA). *What is CHA?* 2006. Online: www.caribbeanhotels.org.

Caribbean Tourism Organization (CTO). 2005. CTO & CHA sign historic coop agreement. *CTO News* 5 (8).

———. 2006. *Caribbean Tourism Organization—Its history, aims and objectives structure and services.* Online: http://www.onecaribbean.org/information/documentview.php?rowid=401.

Gibbings, W. 2000. Caribbean looks inward to solve regional problems. *The Virtual Institute of Caribbean Studies Newsletter* (4) 2. Online: http://pw1.netcom.com/~hhenke/news11.htm.

Girvan, N. 1999. *Globalisation, fragmentation and integration: A Caribbean perspective.* Online: http://www.geocities.com/CollegePark/Library/3954/tourisma.pdf.

———. 2000. *Notes on CARICOM, the ACS and Caribbean survival.* Online: http://www.geocities.com/CollegePark/Library/3954/caribbeansurvival.pdf.

Government of St. Kitts and Nevis (GOSKN). 2001. *Caribbean heads and tourism ministers search for solutions to sustain critical tourism sector.* Online: http://www.stkittsnevis.net/media.html .

Gray, B. 1989. *Collaborating: Finding common ground for multi-party problems.* San Francisco: Jossey-Bass.

———. 1996. Cross-sectoral partners: Collaborative alliances: Business, government and communities. In *Creating collaborative advantage.* C. Huxham, ed. London: Sage.

Griffith, I. L. 2000. *Caribbean geopolitics and geonarcotics: New dynamics, same old dilemna.* Online: http://www.geocities.com/CollegePark/Library/3954/griffi.htm.

Grugel, J. 1995. *Politics and development in the Caribbean basin: Central America and the Caribbean in the new world order.* London: Macmillan.

Hall, C. M. 1994. *Tourism and politics: Policy, power and place.* London: Wiley.

———. 2000. *Tourism planning: Policies, processes and relationships.* Harlow, UK: Prentice Hall.

Holder, J. 2001. Meeting the Challenge of Change. Online: http://www.caricom.org/jsp/press releases/pres149 01.htm.

Jamal, T., and D. Getz. 1995. Collaboration theory and community tourism planning. *Annals of Tourism Research* 22 (1):186–204.

———. 1999. Community roundtables for tourism-related conflicts: The dialects of consensus and process structures. *Journal of Sustainable Tourism* 7 (3 & 4):290–313.

Jessop, D. 1999. No regional unity, no future. *The Virtual Institute of Caribbean Studies Newsletter* (3), 4. Online: http://pw1.netcom.com/~hhenke/news9.htm.

Jordan, L. 2004. Institutional arrangements for tourism in small twin-island states of the Caribbean. In *Tourism in the Caribbean: Tends, development, prospects.* D. Duval, ed. London: Routledge.

Ladkin, A., and A. M. Bertramini. 2002. Collaborative tourism planning: A case of Cusco, Peru. *Current Issues in Tourism* 5 (2):71–93.

Lowenthal, D., and C. G. Clarke. 1980. Island orphans: Barbuda and the rest. *Journal of Commonwealth and Comparative Politics* 18(3):293–307.

Mackintosh, M. 1992. Partnership: Issues of policy and negotiation. *Local Economy* 7 (3):210–24.

Meighoo, K. 2000. *An outmoded notion of sovereignty.* Online: http://www.geocities.com/CollegePark/Library/3954/sovereig.pdf.

Morrison, D. April 4, 2006. Region no longer pivotal. *Trinidad Express,* 11.

Organisation of Eastern Caribbean States (OECS). 2006. Welcome to the Organisation of Eastern Caribbean States. Online: http://www.oecs.org.

Pantin, D. A. 1995. *Finding the "safe havens" for sustainable Caribbean development in the 21st century.* Online: http://www.tidco.co.tt/local/seduweb/research/dennisp/martniq.htm.

Payne, A., and P. Sutton. 2001. *Charting Caribbean development.* London: Macmillan Education Ltd.

Reed, M. 2000. Collaborative Tourism planning as adaptive experiments in emergent tourism settings. In *Tourism collaboration and partnerships: Politics, practice and sustainability.* B. Bramwell and B. Lane, eds. Clevedon, UK: Channel View.

Richardson, B. C. 1992. *The Caribbean in the wider world, 1492–1992.* Cambridge: Cambridge University Press.

Selwyn, P. 1975. Introduction: Room for manoeuvre? In *Development policy in small countries.* P. Selwyn, ed. London: Croom Helm.

St. Kitts-Nevis. 1998. Constitutional Commission, Report of a Constitutional Commission appointed by His Excellency The Governor General of Saint Kitts and Nevis. London: Constitutional Commission.

Timothy, D. J. 1998. Cooperative tourism planning in a developing destination. *Journal of Sustainable Tourism* 6 (1):52–68.

———. 2004. Tourism and supranationalism in the Caribbean. In *Tourism in the Caribbean: Tends, development, prospects.* D. Duval, ed. London: Routledge.

United Nations (UN). 2006. *Partnerships for sustainable development.* Online: http://www.un.org./esa/sustdev/partnerships/partnerships.htm.

Watson, J. B. 1979. *The West Indian heritage: A history of the West Indies.* London: John Murray Publishers Ltd.

Williams, E. 1970. *From Columbus to Castro: The history of the Caribbean 1492–1969.* London: Andre Deutsch Limited.

12 New Directions in Caribbean Tourism Education
Awakening the Silent Voices

Acolla Lewis

INTRODUCTION

The regional development of tourism has given rise to increased employment opportunities in the field and, along with that increase, a need for better means of educating the industry's workforce. In recognition of the need for people educated in tourism operations, governments in the Caribbean have sent students overseas to centres primarily in the United Kingdom. However, the focus of most of these programmes has been on courses designed to prepare graduates for employment in the tourism industry in developed mainland tourist destinations, that is, in contrast to the tourism environment in the region. Chambers (1997) observed that, in many cases, programs developed in North America and in the United Kingdom have been dominated by economic models and by a vocational ethos that placed minimal emphasis on a critical assessment of the varied consequences of tourism.

In response to the rapid rise in tourism demand in the Caribbean and the acknowledgment of a need for tourism education, a number of institutes for tourism education were established throughout the Caribbean (Charles 1997). In the establishment of many of these institutes, Theuns and Go (1992) note that "Western" models have been imported without taking sufficient account of the needs of the local tourism sector and the existing social, cultural, and economic framework of the host country. Furthermore, the tourism educators at these institutes were educated in the "international classroom" (in North America and the United Kingdom), and thus the tourism knowledge constructed in the region is primarily Eurocentric.

The governments and leaders of the Caribbean have finally and almost unanimously come to the view that tourism is anything from "an important," to "the most important," to "the only" means of economic survival for their states (Pattullo 1996). The extent of the Caribbean's dependence on tourism is manifested in the fact that tourism earnings account for approximately 25 percent of the region's gross domestic product (GDP), and the sector provides employment for 25 percent of the region's labour market (Hall et al. 2005). This dependence on tourism is exacerbated by the difficulties experienced regionally in the agriculture sector as well as in the exportation of manufactured products such as textiles. Given the growing importance of tourism to the economic development of the Caribbean islands, and the constantly changing internal and external environments, "it seems timely to put Caribbean tourism under a microscope, to create a new vision of what it should be, what should be its developmental objectives, and what strategies are needed for achieving them" (Hall et al. 2005:7).

An integral component of this critical review of Caribbean tourism is a reassessment of the approach to tourism education and training in the region. More specifically, the question needs to be asked, how can tourism education respond to the present realities shaping Caribbean tourism? In a review of the bachelor of science in tourism management at the Centre for Hotel and Tourism Management in the Bahamas conducted by the Office of the Board of Undergraduate Studies OBUS (2000:4), it was indicated that "the existing degree and diploma programmes lack distinctiveness and do not play a major leadership role in tourism and hotel management in the region."

The chapter opens with a discussion on tourism education in the international classroom and its influence on Caribbean tourism education. It thus critically examines the impact of Eurocentric tourism knowledge on Caribbean tourism education. This is followed by an examination of the challenges confronting the tourism industry in the region. Here, it will address the issue of how tourism education can be reformulated in such a way that the previously silent voices can contribute to the creation of "new" tourism knowledge for the region. The chapter concludes with the proposal that a stakeholder informed approach to tourism curriculum development is the way forward for tourism education in the region.

TOURISM EDUCATION IN THE INTERNATIONAL CLASSROOM

Although there is evidence of increasing diversity in tourism degrees (Tribe 2002), tourism curriculum development in the international classroom (UK and North America) has mainly been of the vocational type. A study by Airey and Johnson (1999) on the profile of tourism degrees in the United Kingdom attests to this. They surveyed, summarised, and ranked the stated aims and objectives of tourism degrees in the United Kingdom. The summary of aims and objectives indicate a mixture of both vocational and liberal aims. However, the ranking demonstrates clearly the extent to which tourism degrees in the United Kingdom are operating within narrow dimensions.

They discovered that the aims and objectives are dominated by operational and vocational concerns. More than three-quarters of all courses stress "career opportunities" and more than half give attention to "employment/ employer links." Only three of the twenty aims offer any analysis of the wider tourism world beyond the vocational element, and they all appear in the lower part of the ranked list of aims and account for only 11 percent of the totals mentioned. These are "sound education/academic understanding," "broad foundation/wide range/thorough grounding," and "social context/ sustainable tourism." Stuart (2002), in more recent research on undergraduate tourism degrees in the United Kingdom, reveals that operational aims still dominate the stated aims of the majority of tourism degrees.

According to Cooper et al. (1994), the apparent dominance of a vocational emphasis in tourism curriculum development is closely tied to the traditions of tourism education. Baum (2001) observed that the main thrust of tourism education commenced and remains within the vocational-education sector. Cooper et al. (1994) suggest that it is possible to discern two distinct ways in which the study of tourism has developed as an academic endeavour. First, sector-based vocational courses for the travel trade have been developed and have had a strong influence on the direction of tourism education and training. Cooper and Shepherd (1997) explain that training has traditionally dominated the tourism sector where vocationally oriented courses have provided the workforce with the necessary craft skills for many years.

Second, tourism courses have developed as a means of enriching business studies by giving them a vocational orientation. This is evidenced by the report on the profile of tourism degree courses in the United Kingdom published by The National Liaison Group for Higher Education in Tourism (NLG; Airey and Johnson 1998). They noted that of the ninety-nine tourism courses surveyed, around one-third are located in departments of business and management. They suggest that this demonstrates the continuing emphasis placed upon vocational courses with a business-management orientation.

It is possible to attribute this development of tourism curriculum with a vocational emphasis to several key factors, notably the demands of industry for a more professional workforce. There is a growing recognition in the tourism industry that a vocationally trained workforce is a key element in maintaining competitive advantage (Airey 1999). Cooper and Shepherd (1997) noted that research undertaken by the University of Surrey in 1991 suggests that larger tourism enterprises are placing more emphasis on both training and education as companies are understanding more fully the role of training and education in boosting company profitability, productivity, and turnover. With this increasing positive view towards training and education, industry is demanding a more professional workforce. This has resulted in the continued proliferation of vocational courses to the extent that they have expanded to embrace a much wider set of industrial sectors (Cooper and Shepherd 1997; Morgan 2004).

Caribbean Tourism Education

It is this vocational ethos that has dominated the development of tourism curricula in the region. The majority of educational contribution to the industry is at the tertiary level, and the limited education at the primary and secondary levels are confined primarily to the social studies curriculum, which is not consistent between countries. Seward and Spinrad (1982) observed that the decade of the sixties was characterized by great optimism toward the potential for a viable tourism industry in the region. In the initial rush to capitalize on rapid expansion of tourist demand, local governments therefore provided generous financial incentives to attract foreign investors willing to develop hotels and related projects. It was during this time that the governments of Barbados and Jamaica, two of the more developed tourist destinations, saw the need to train their nationals for entry-level positions in hotels. To this end, the Barbados Hotel School was established in 1966 as a culinary school. The Jamaica Hotel School started its operations soon after in 1968 with a focus on equipping individuals with skills in the front office, housekeeping, dining room, and kitchen departments (O'Reilly 2002). These two hotel schools were the prototype for other schools in the region such as the Trinidad and Tobago Hotel School (1970s) and the Bahamas Hotel Training College (1975).

As a number of destinations began to approach the development stage in their life cycle in the 1970s, there was an urgent need for the education and training of local managers for the regional hotel and tourism industry. In response to this need, the University of the West Indies (UWI), Department of Management Studies established in Jamaica a hotel-management degree programme in 1977. UWI's degree programme was designed to educate, train, and generally prepare Caribbean students to assume future leadership roles in the region's hospitality and tourism industry. As the name suggests, the programme focused specifically on hotel management, and thus there was limited or no attention given to the wider socioeconomic and cultural issues arising from tourism development in the region. It was not until 1983 that a bachelor's degree in tourism management was introduced to address some of these key issues. Both programmes were then transferred to Nassau, Bahamas, where the Centre for Hotel and Tourism Management (CHTM) was established, as this island was viewed as having the best-developed tourism plant for effective university-industry interface.

From their inception, the tourism and hospitality programmes have received teaching, research, consultancy, and curriculum-development assistance from hospitality and tourism programmes both in the United Kingdom and North America, such as the University of Surrey in England and Florida International University in North America. In developing the tourism curriculum, there was a need for some benchmark, and so the curriculum planners turned to the more established and recognised programmes in the United Kingdom and North America. As such, the key objectives of the curriculum were, and still are to a great extent, focused on the vocational element. One of the main objectives identified in the inception of the programme was the training of students in certain specific skills to prepare them to assume management responsibilities and equip them to perform certain specific functions in the various sectors of the industry (O'Reilly and Charles 1990).

More recently, there has been the introduction of four-year joint bachelor of science degrees in hospitality and tourism management between the

UWI and tertiary institutions throughout the region. These associations are referred to as a 2 + 2 arrangement whereby students complete two years at a partnering institution and two years at any one of the three UWI campuses. In general, the 2 + 2 programme is aimed at producing graduates who have

- Immediate functional competencies;
- An entrepreneurial and innovative attitude to work;
- A good overall knowledge of the business and competitive environments in the industry;
- Strong people skills;
- Language and cultural sophistication, awareness, and appreciation;
- A strong sense of social obligations and environmental responsibilities of the industry;
- Respect for vocational work (Jayawardena and Cooke 2002:53).

The aims of the curriculum clearly emphasise the business and technical skills needed for a career in the industry. Minimal attention is given to the liberal aspect of the curriculum, where the wider issues in the society become paramount. This has resulted primarily because tourism curricula have been informed by two main stakeholders—industry and academics. For the former, tourism knowledge is selected insofar as it prepares graduates to contribute to the efficiency of tourism services, thereby enhancing the industry's profitability. Academics bring their own interests, expertise, and prejudices to the construction of tourism knowledge. The majority of tourism academics in the Caribbean have either been schooled in the international classroom and/or at the CHTM, which was established on a Eurocentric foundation.

In considering the tourism curriculum in Caribbean islands, it is the wider sociocultural, political, and economic contexts that take centre stage. Given the critical importance of tourism to these island economies, tourism educators are faced with the challenge of ensuring that the curriculum prepares students to plan, manage, and develop tourism in the islands, as well as responding to the key global and local issues that affect the wider society. In other words, attention must be placed on how the sociocultural, political, and economic issues can be reflected in the vocational and liberal agenda of the tourism curriculum. This can only be achieved with collaboration with key stakeholders.

Caribbean Tourism Realities

In the words of Hegarty (1990:41),

> Educational programs for development must be placed in the context of their societies' needs and be interconnected with well conceived development plans. . . . The onus is therefore placed on the hospitality and tourism education programs to demonstrate that knowledge

benefits not only the individual who acquires it, but also the society of which he or she is a part.

In agreement, Smith and Cooper (1999) noted that the curriculum must be context related but not context bound. It is this notion of a context-related curriculum that demands a rethink of tourism education in the region and a refocusing on its role in a changing global and regional tourism environment.

Caribbean tourism can be defined according to four main realities. First, tourism is the lifeblood of the majority of Caribbean economies. In the words of Mather and Todd (1993:11), "There is probably no other region in the world in which tourism as a source of income, employment, hard currency earnings and economic growth has greater importance than in the Caribbean." In the Caribbean, the share of tourism in GDP ranges from a third to a half for most islands. In employment terms, a recent report by the World Travel and Tourism Council (WTTC) reveals that Caribbean employment in tourism is estimated at 2,643,000 or 1 in every 6.5 jobs (15.4% of total employment). This is forecasted to rise to 17.1 percent by 2014 and the industry is expected to generate US$51.3 billion in economic activity in 2006 and account for 16.4 percent of GDP (WTTC 2004).

Second, the industry is largely dominated by foreign transnationals. In the region where tourism is the leading economic sector, foreign ownership, high leakage, and expatriate domination of management are at high levels (Hall and Page 1996). In most of the Caribbean, the level of leakages averages around 70 percent, which means that for every dollar earned in foreign exchange, 70 percent is lost in imports. Foreign-exchange earnings are lost by way of payments for exports, expatriate staff salaries, profit repatriation of transnationals, and rising consumption by locals of imported goods made available through the industry (Patullo 1996). The foreign dominance of the industry in the region raises the issue of power and control. The concern for the islands is that the locus of control over the tourism development process is not in the hands of the people that are most affected by development, the host community, but in the hands of the tourist-generating regions.

Third, the local cultures in the islands are increasingly influenced by Western culture. The underlying concern here is the threat of cultural dependency whereby the islands' beliefs and values appear to be determined by North America and Western Europe to the neglect of their own identity. Much of what is admired within the Caribbean and is seen to be "better" remains foreign (usually North American), whether in design, technology, food, or the visual arts. With the continued increase in tourist demand to the islands, the leaders of the tourism industry, both foreign and local, have seemed generally unconcerned to protect the authenticity of Caribbean art forms, dance forms, and heritage. The problem of a lack of self-confidence and a lack of pride in one's identity is a deep-seated problem in Caribbean culture (Lewis 2002).

Fourth, the traditional consumer-perceived generic product of the Caribbean is crystallised in the phrase *escape the winter to tropical sun, sea and sand*. The Caribbean islands are in a general sense tourist-receiving destinations supplying primarily hospitality-based services including accommodation and catering facilities, infrastructural, and other destination-specific facilities. The Caribbean tourism product has come under intense criticism because it has been deemed outdated, poorly managed, and in need of upgrading (Hall et al. 2005). The region is further characterized by relatively high travel costs and room rates, high labour and utility costs, and low profits and income multipliers (Tewarie 2002). To maintain its position in the global tourist economy and to develop in a sustainable manner, the Caribbean will have to devise alternative strategies to traditional mass tourism promotion that emphasise quality over quantity.

According to Botterill and Gale (2005:478), "Western academics have been the architects of tourism knowledge" in the international classroom. As such, the knowledge created is "situated knowledge," which only marginally encompasses the tourism realities confronting the developing world. Therefore, it is with this understanding of the economic and sociocultural environment in the Caribbean that tourism educators are challenged to develop timely and relevant tourism curricula that are context related and not context bound. In the words of Botterill and Gale (2005:478), "Previously silent voices [need to] contribute to create 'new' tourism knowledge."

NEW DIRECTIONS IN CARIBBEAN TOURISM EDUCATION

The provision of high-quality education and training is the most critical component in the future growth and development of the tourism industry in the Caribbean. Over the years, tourism education and training have come under intense criticism with respect to the governance and operation of the institutions as well as the quality of the programmes. Conlin and Titcombe (1995:67) characterized education in the region as "fragmented, unco-ordinated, and occasionally redundant." The lack of integration in the approach to tourism education has resulted in a flood of external education agencies promoting their various products in the region, with no consideration of the critical issues discussed previously that challenge Caribbean economies, that tourism education need to address. It has now become necessary to alter the philosophical foundation of vocationalism of tertiary tourism education in the region to embrace the wider issues that affect regional economies. Moreover, any country or region competing in the international tourism market today requires a well-developed tourism-education strategy. It is on this premise that a stakeholder-informed approach to tourism education is proposed as a useful framework for the planning and developing of national tourism-education strategies for Caribbean destinations.

Stakeholder-Informed Approach

The stakeholder approach focuses on facilitation of useful interventions and recognizes the widely divergent, yet equally significant, worldviews of different stakeholders. This approach advocates consultation with a wide array of individuals in the local society who are influenced by and who can influence the direction of tourism-education decisions at the destination. The rationale here is that the different groups in the local society indicate the values and needs of the society; they speak for the culture of the destinations, and they reveal the general attitude of the locals towards tourism development. Consultation with this broad cross section of individuals allows tourism educators in Caribbean territories to make better-informed tourism curriculum decisions that respond to the interests and needs of the tourism industry and the wider society. Such a perspective involves learning about existing realities and making appropriate choices in extremely complex and divergent situations. This approach is an inclusive and participatory process that is guided by stakeholder values and underlying philosophical underpinnings where all stakeholders stand to benefit and where tourism education is less likely to be influenced by ideological biases. See Figure 12.1 for a framework of the proposed approach.

This approach involves a number of steps, which begins with the systematic identification of the various stakeholders, their interests, and interrelationships

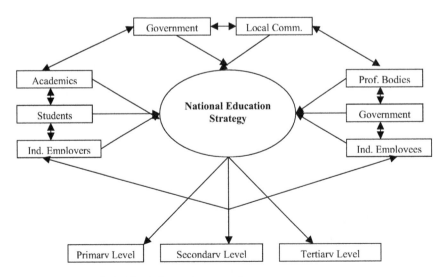

Figure 12.1　Stakeholder-informed approach.

as far as tourism education is concerned. According to Cooper and Westlake (1998) and Lewis (2006), the two groups of primary stakeholders considered to be at the core of tourism education are the academics and the students. The academics are seen as the producers of the educational experience in that they are ultimately responsible for the planning, development, and delivery of the tourism curriculum. The students are the direct consumers of tourism education because they are the primary recipients of tourism education. The other main groups identified in the research that should be consulted include industry employers, industry employees, government, local community, and professional bodies.

The second step involves the gathering of stakeholder views through personal interviews and focus groups. The focus of this stakeholder consultation is on the eliciting of generic developmental factors that are affecting the different Caribbean territories, and those factors that relate more specifically to the structure, form, and content of tourism education. In research conducted by Lewis (2006) in three Caribbean territories (Bahamas, Barbados, and Jamaica), representatives from the stakeholders groups identified previously were asked to share from a personal perspective their views on the main priorities with tourism development in the respective islands. The areas delineated include foreign dominance of the industry in terms of the management of tourism organisations, a lack of professional employees with the necessary knowledge and skills for the industry, and a lack of planning of the development of tourism resulting in significant negative implications for the natural environment. These contextual factors provide a broad local context for the respective destinations that tourism curricula must closely relate to if they are to have socioeconomic relevancy and cultural value.

The views of the stakeholders are integral in informing the third step, the identification of the strategic priorities for tourism education in the respective destinations. Importantly, the strategic priorities must be established with cognizance of both the national development plans and the national tourism development policy. The strategic priorities should be the outcome of a consideration of the changes in the global economy that impact tourism development in the region, the respective national tourism agenda, and the input of the relevant stakeholders. These priorities should inform the development of an inclusive national strategy for tourism education. The aim of a tourism-education strategy should be to provide the tourism industry with a functional and effective workforce, thereby contributing to the efficiency and profitability of tourism businesses. These individuals should also have a holistic understanding of tourism in the islands and can contribute to the overall betterment of the tourism society.

The final and most challenging step in this approach is the development of a strategy for implementation at the various levels of the education system. The approach to tourism education in the region has been criticized for its lack of integration, and this is evident in the absence of a comprehensive tourism-education agenda for the different levels of educational institutions.

Since tourism is pivotal to the Caribbean's economic survival, it is critical that issues on how to coordinate and integrate different levels of education and other forms of professional curricula are considered, in order to allow for continuity and systematic progression from one stage of learning to another. There has been a commendable attempt at integration at the tertiary level in the creation of the Association of Caribbean Tertiary Institutions (ACTI) and the Caribbean Tourism Human Resource Council (CTHRC). The latter represents a critical milestone in the development and coordination of tourism education in the region as the council continues to support tourism human resource development through its annual Tourism Educators Forum, the Scholarship Foundation Programme, and, more recently, the establishment of a Caribbean Tourism Learning System (CTLS). Some of the key elements of the CTLS include a unified core curriculum for different levels of certification, a system for transfer of credits between institutions, and student exchange programmes (Lewis 2005). Further, vertical articulation is needed at both the primary and secondary levels.

Application of the Stakeholder-Informed Approach

In June 2006, the University of the West Indies (UWI), St. Augustine Campus, adopted a stakeholder-informed approach in the development of its master of science in tourism. Over a one-month period, 150 questionnaires were distributed throughout the Caribbean to representatives from six stakeholder groups including academics, students (past and present), industry employers, industry employees, government representatives, and professional bodies. The questionnaire focused on two core areas: the main challenges facing the hospitality and tourism industry in the region and the primary area of focus for the proposed master of science programme.

While several challenges facing the industry were raised, the core issues noted were:

- Unplanned and inappropriate developments;
- Lack of development and implementation of policies;
- Inadequate resources for research and planning for management of industry;
- No enforcement of sustainable tourism practices.

With a focus on such issues, it was not surprising that the majority of participants indicated that the core areas that the proposed master of science should emphasise include sustainable development and tourism planning and policymaking. Subsequently, the Department of Management Studies developed a master of science in tourism development and management with an emphasis on the areas identified.

Significant to note from the stakeholder exercise was the extent to which the focus of the postgraduate programme was linked to the challenges facing the hospitality and tourism industry. At the core of the stakeholder approach is the facilitation of curriculum decisions that address the needs of the specific society. Stakeholder involvement ensured that the programme development was not purely an academic exercise that catered to the interests of the academics but one that was outward looking, thereby embracing the interests of those that can influence and who are influenced by curriculum decisions.

Concerns with the application of the stakeholder approach

Although there are pertinent benefits to be gained from the adoption of the stakeholder approach, there are two noteworthy concerns with the practical application of the approach in tourism-education decision making in the Caribbean that need to be considered. The first of these concerns is the area of stakeholder power. Freeman (1984), the advocate for stakeholder theory, alerted us to the issue of dealing with stakeholder power and competing interests among stakeholders. One of the expected outcomes of stakeholder involvement in decision making is the emergence of a number of conflicting interests among stakeholders that warrant careful manipulation on the part of decision makers. For this reason, Freeman (1984) stresses that decision makers should avoid attempting to satisfy the needs of all the stakeholders. Instead, the idea of applying stakeholder theory to tourism-education decision making is to capture the interests of stakeholder groups across the power continuum at the instructional, institutional, and societal levels.

In the development of the postgraduate programme at the UWI, the issue of stakeholder power manifested itself in the development of the course offerings. Clearly, in the final analysis it is the academics that determine the programme content and delivery. The challenge the UWI encountered was the ability of the academics to deliver what has been established by the stakeholder exercise without being prejudiced by their individual biases.

A further concern regarding the practical application of the stakeholder approach is what Freeman (1984) refers to as analysis paralysis. Stakeholder inclusion in decision making is a lengthy, time-consuming, expensive exercise that can dissuade educators from embarking on this approach. However, a lack of stakeholder involvement can result in a partially framed curriculum that favours only the profit-seeking interests of tourism organisations and the employment interests of students (Tribe 2002). In conducting the stakeholder exercise for the master of science programme, it was noted that the use of questionnaires proved to be more effective in terms of shortening the length of the exercise.

CONCLUSION

The primary focus of tourism education in the United Kingdom and North America in particular is on preparing students for employment in the tourism industry (Airey and Johnson 1999). It can be argued that the principal reason for this lies in the fact that tourism is viewed as an industry to be managed in order to generate benefits for the profit-seeking tourism organisations, the pleasure-seeking tourists ,and to further perpetuate capitalists' interests. From a Caribbean perspective, tourism education has to respond to much broader sociocultural, economic, and environmental issues. The context of tourism development cannot be divorced from the overall purposes of tourism education. Thus, Caribbean tourism education should seek to satisfy the tourism labour market, to satisfy the tourists' wants, and to promote economic welfare.

The proposed stakeholder-informed approach is an appropriate and systematic method for ensuring that the plurality of interests to be served by tourism education is adequately captured. Furthermore, the development of a national tourism-education strategy through stakeholder input addresses the perennial challenge of fragmentation in the tourism education system throughout the region. To this end, the application of the stakeholder-informed approach to tourism education is crucial in the development of a dynamic and resilient workforce that is key to shaping the Caribbean tourism society and gaining a competitive advantage in the global tourism market. With this in mind, the knowledge created in the Caribbean classroom must embrace both a vocational and a liberal agenda. More specifically, the former should emphasise the preparation of students for effectiveness at work in the industry. The latter should ensure that students adopt a critical stance towards tourism and develop a practical knowledge of dealing with the key issues regarding tourism development in the islands.

REFERENCES

Airey, D. 1999. Education for tourism—East meets West. *International Journal of Tourism and Hospitality Research* 1 (1):7–18.

Airey, D., and S. Johnson. 1998. *NLG guideline no. 7: The profile of tourism studies degree courses in the UK: 1997/98*. London: National Liaison Group for Higher Education in Tourism.

Airey, D. , and S. Johnson. 1999. The content of degree courses in the UK. *Tourism Management* 20 (2):229–35.

Baum, T. 2001. Education for tourism in a global economy. In *Tourism in the age of globalization*. S. Wahab and C. Cooper, eds. London: Routledge.

Botterill, D., and T. Gale. 2005. Postgraduate and Ph.D. education. In *An international handbook of tourism education*. D. Airey and J. Tribe, eds. London: Elsevier.

Caribbean Tourism Organisation, Working Document for the Regional Summit on Tourism. 2001. Available: http://www.caricom.org/tourismdocuments.htm.

Chambers, E. 1997. Tourism as a subject of higher education: Educating Thailand's tourism workforce. In *Tourism and culture: An applied perspective*. E. Chambers, ed. New York: State University of New York Press.

Charles, K. 1997. Tourism education and training in the Caribbean: Preparing for the 21st century. *Progress in Tourism and Hospitality Research* 3:189–97.

Conlin, M. V., and J. A. Titcombe. 1995. Human resources: A strategic imperative for Caribbean tourism. In *Island tourism: Management principles and practice*. M. V. a. B. Conlin, ed, T. Chichester, UK: Wiley.

Cooper, C., and R. Shepherd. 1997. The relationship between tourism education and the tourism industry: Implications for tourism education. *Tourism Recreation Research* 22 (1):34–47.

Cooper, C., R. Shepherd, and J. Westlake. 1994. *Tourism and hospitality education*. Guildford: University of Surrey.

Cooper, C., and J. Westlake. April 1998. Stakeholders and tourism education. *Industry and Higher Education*.

Freeman, E. 1984. *Strategic management—a stakeholder approach*. London: Pitman.

Hall, K., J. Holder, and C. Jayawardena. 2005. Caribbean tourism and the role of the University of the West Indies. In *Caribbean tourism: Visions, missions and challenges*. C. Jayawardena, ed. Jamaica: Ian Randle Publishers.

Hall, M., and S. Page, eds. 1996. *Tourism in the Pacific: Issues and cases*. London: International Thomson Business Press.

Hegarty, J. 1990. Challenges and opportunities for creating hospitality and tourism education programs in developing countries. *Hospitality and Tourism Educator* 2 (3):40–41.

Jayawardena, C., and M. T. Cooke. 2002. Challenges in implementing tourism and hospitality joint degrees: The case of the University of the West Indies and University of Technology, Jamaica. In *Tourism and hospitality education and training in the Caribbean*. C. Jayawardena, ed. Jamaica: University of the West Indies Press.

Lewis, A. 2002. *A case study of tourism curriculum development in the Caribbean: A stakeholder perspective*. Unpublished doctoral dissertation, Brunel University, UK.

———. 2005. The Caribbean. In *An international handbook of tourism education*. A. D. and J. Tribe, eds. London: Elsevier.

———. 2006. Stakeholder informed tourism education: Voices from the Caribbean. *Journal of Hospitality, Leisure, Sport and Tourism Education* 5 (2):14–24.

Mather, S., and G. Todd. 1993. Tourism in the Caribbean (Special Report No. 455). London: Economist Intelligence Unit.

Morgan, M. 2004. From production line to drama school: Higher education for the future of tourism. *International Journal of Contemporary Hospitality Management* 16 (2):91–99.

Office for the Board of Undergraduate Studies (OBUS), Review of the Centre for Hotel and Tourism Management, Nassau, Bahamas. 2000. Unpublished document, Office for the Board of Undergraduate Studies. 2000. Mona, Jamaica.

O'Reilly, A. M. 2002. Past, present and future of tourism and hospitality education in the West Indies. In *Tourism and hospitality education and training in the Caribbean*. C. Jayawardena, ed. Jamaica: University of the West Indies Press.

O'Reilly, A. M., and K. R. Charles. 1990. Creating a hotel and tourism management programme in a developing country: The case of the University of the West Indies. *Hospitality and Tourism Educator* 18:47–49.

Pattullo, P. 1996. *Last resorts: The cost of tourism in the Caribbean*. London: Cassell.

Seward, S., and B. Spinrad. 1982. *Tourism in the Caribbean: The economic impact*. Ottawa: International Development Research Centre.

Smith, G., and C. Cooper. 1999. *The challenge of globalisation from theory to practice: Competitive approaches to tourism and hospitality curriculum design: The case of the International School of Tourism and Hotel Management, Puerto Rico*. Paper read at Latin American Tourism in the Next Millennium: Education, Investment and Sustainability, at Panama City, Panama,.

Stuart, M. 2002. Critical influences on tourism as a subject in UK higher education: Lecturer perspectives. *Journal of Hospitality, Leisure, Sport and Tourism Education* 1 (1):5–18.

Tewarie, B. 2002. The development of a sustainable tourism sector in the Caribbean. In *Island tourism and sustainable development: Caribbean, Pacific and Mediterranean experiences*. A. Apostolopoulos and D. J. Gayle, eds. London: Praeger.

Theuns, H. L., and F. Go. 1992. need led priorities in hospitality education for the Third World. *World Travel and Tourism Review* 2:293–302.

Tribe, J. 2002. The philosophic practitioner. *Annals of Tourism Research* 29 (2):338–57.

World Travel and Tourism Council, The Caribbean: 2004. *The impact of travel and tourism on jobs and the economy.* London: WTTC.

13 Towards New Ontologies of Caribbean Tourism

Marcella Daye, Donna Chambers, and Sherma Roberts

... visitors to the Caribbean must feel that they are inhabiting a succession of postcards. For tourists, the sunshine cannot be serious ... in the unending summer of the tropics not even poverty or poetry ... seems capable of being profound because the nature around it is so exultant, so resolutely ecstatic, like its music. A culture based on joy is bound to be shallow. Sadly, to sell itself, the Caribbean encourages the delights of mindlessness, of brilliant vacuity, as a place to flee not only winter but that seriousness that comes only out of culture with four seasons ...
(Excerpts from Derek Walcott, Nobel Lecture, December 7, 1992)

This quotation from Caribbean poet laureate Derek Walcott, although written a decade and a half ago, still reflects one of the key challenges facing the Caribbean tourism industry and which opened the discussion in this volume. This key challenge centres on the way that the Caribbean is represented and imagined by the Western tourist and the extent to which the Caribbean is complicit in this touristic representation and imagining of itself. Indeed, the way in which the Caribbean is represented in and through the tourism industry (which, as many of the chapters in this volume have indicated, is the main foreign-exchange earner for most of the economies in the region) has assumed increasing importance in what has become a fiercely competitive global tourism environment. The exoticised image of the Caribbean as paradise and as a ludic space for the Western traveller has ostensibly served the region's tourism industry for several decades. But in the twenty-first century, with more and more countries entering the tourism arena, coupled with the change in consumer tastes towards more varied holiday experiences, the Caribbean, as a mature destination area, needs to seriously rethink its image and brand strategies in order to ensure differentiation, and by extension, sustainable competitive advantage.

In this regard, Daye, in the opening chapter to this volume, advocated the need for a disruption of colonial stereotypical images of the Caribbean through a 're-visioning' of Caribbean tourism to reflect the processes of indigenisation and creolisation. What this means is that in reimaging the Caribbean brand, tourism marketers in and of the Caribbean need to draw on the diverse, rich cultural inheritances of the region, which are to be found in the music, food, dance, folk tales, inter alia, all of which have been traditionally submerged under normative and homogenising discourses of the Caribbean as Paradise. Such a privileging of indigenous cultural forms and expressions would, Daye argues, be integral to a differentiation strategy for the region's tourism sector. We argue further here that a focus on processes of indigenisation and creolisation should not take place only within the tourism space but should pervade every aspect of Caribbean life and society. It is only then that the region can claim an identity of its own which is not simply a reflection of the tourist gaze.

While, admittedly, there are certain *problematiques* associated with processes of indigenisation and creolisation, as noted in the introductory chapter to this volume, these processes are nevertheless pertinent in a Caribbean context. The other chapters of this volume also implicitly continue this discussion by suggesting that Caribbean countries need to seriously consider the explicit fostering and promotion of novel forms of tourism, which include indigenous and cultural resources such as music and food. Indeed, like music, food is a very important expression of what it means to be 'Caribbean,' and Dodman and Rhiney have told the story of how food is produced and consumed in the tourist space of Negril, Jamaica. While looking at the important economic linkages (between tourism and agriculture) that this niche area provides, the authors further consider the way food also serves to reinforce cultural identity. Indeed, the importance of food to Caribbean tourism experiences is noted by Hartley (2007), who asserts that 'Now people are visiting islands such as Grenada, Barbados, Anguilla and St. Barts, an epicurean outpost in the French Antilles, because of the food, not in spite of it. Even those with slim wallets can enjoy some of the best beach bars and spiciest street food this side of South East Asia.'

This serves to emphasise the inextricable link between tourism and the cultural identity of the islands, which begs further exploration.

For his part, Webster has pointed to an already existing, though largely unrecognised, microniche market represented by the vinyl tourist, who engages in tourism ostensibly in order to 'plunder' the tangible cultural heritage of Jamaica (the vinyl record). However, Webster questions whether this microniche form of tourism is sustainable given the exhaustive nature of the supply and also, importantly, he raises ethical questions associated with this 'cultural plunder' of invaluable collections of Jamaican music. Implicated in this discussion is the range of the region's cultural assets that is being covertly appropriated by tourists from the developed world and the extent to which this might represent a new and insidious form of cultural exploitation.

With indicators of a slowdown in tourist arrivals to the region, the quest for achieving sustainable competitive advantage for the Caribbean tourism industry is a relevant concern underpinning many of the chapters in this volume. However, we recognise that there is also a need to strengthen the internal institutional structures of the region by focusing on the further development of regional tourism partnerships as discussed by Jordan in this volume. There are already recent indications that regional tourism partnerships can achieve some measure of success. Specifically, during the process of putting together this edited volume the Cricket World Cup 2007 took place and can be seen as an example of a regional partnership which was to some extent successful. For forty-seven days, eight Caribbean countries came together to host one of the largest global sporting events, and this is perhaps testimony to the fact that partnerships can deliver some benefits. That is, this event required considerable cooperation from regional governments in the areas of safety and security, movement of people, transportation, technology, human resources, and technical skills.

Although the economic benefits are still unclear, perhaps because of overstating projected arrivals and by extension tourist expenditure, the 2007 Cricket World Cup has certainly left a legacy of integration among the governments of the Caribbean community. The single domestic space is being extended, so that nationals and nonnationals can move with greater ease and less restrictions; the security infrastructure is being maintained and upgraded, and there have been recent discussions about a Caribbean ID card, which would allow Caribbean Common Market nationals to move more freely within the territories. Importantly, the gains of collaboration during the World Cup have strengthened the movement towards the CSM (Caribbean Single Market). However, while these are clearly steps in the right direction, further investigation will be required into some of the more long-term institutional, political, and indeed socioeconomic impacts of the 2007 Cricket World Cup.

Continuing the theme of internal institutional strengthening, Lewis speaks to the need for the Caribbean to adopt a stakeholder-informed approach to tourism education. This approach would take into account the peculiarities of Caribbean politics, economies, and social structures, allowing for these to be infused into the Caribbean tourism curricula. In this regard, there have been some nascent attempts to 'indigenise' the Caribbean tourism curricula. The Cave Hill Campus of the University of the West Indies (UWI) in Barbados now has a master of science in tourism and hospitality management, the curriculum of which has informally been influenced by regional industry practitioners. While the courses are still somewhat similar to those taught in North America and the United Kingdom, there is a very strong Caribbean focus but also inclusive of global tourism trends and international case studies. More recently, as Lewis indicates, the St. Augustine campus of the UWI in Trinidad has, through the use of questionnaires, sought stakeholder input into the tourism and hospitality education needs of the Trinidad and Tobago industry. However, the importance of the provision of postgraduate tourism training across the region is demonstrated with the other two UWI campuses in Jamaica and the Bahamas also delivering similar courses. Undoubtedly there is much more work to be done in this area as the Caribbean tourism industry seeks to improve its product, service delivery, managerial competencies as well as policy and planning by strengthening its human resource capacity.

Many of the chapters, while implicitly or explicitly supporting indigenisation and creolisation, nevertheless recognise and advocate the need for internal *cultural* change if the tourism industry is to address other key challenges, such as that of HIV/AIDS, which Grenade has indicated require, amongst other things, a cultural revolution to address behaviour. Chambers has similarly alluded to the need for such a revolution this time in attitudes towards homosexuality and, by extension, gay travel. However, especially in regard to the latter, the argument is that, given the colonial legacy of the island of Jamaica, in order to facilitate a cultural shift in homophobic attitudes, there needs to be a process of education and dialogue rather than one of coercion by the former colonial power through economic 'sanctions.' This cultural shift is vital for the creation of gay tourism space and the consequent ability of the island to compete in a globalised world in which the gay consumer is seen as one of the most lucrative market segments.

Within the Caribbean itself it is recognised that there are exclusionary discourses and practices which affect the tourism industry and which have received limited attention. Brooks, in her discussion of sociospatial exclusion in tourism, advocates the need to reconfigure the tourism space to include the informal sector of small traders and local craft vendors. Roberts speaks to the necessity for more flexibility in the discourse of sustainable tourism development which takes account of the political and economic realities of small businesses in marginalised Caribbean countries such as Tobago. Ramcharitar discusses the way in which tourism (through cultural events such as carnival) is used to promote and reinforce an exclusionary nationalist discourse in Trinidad which privileges Afro-Trinidadians and silences Indo-Trinidadians, despite the fact that the latter constitute about half of the country's population. The point of emphasis is that there are indicators that within the Caribbean, tourism may be co-opted for the development and nurturing of exclusionary discourses and practices.

Of course, there are many exogenous factors which will have an impact on the sustainable competitive advantage of Caribbean tourism and over which the region has little control. Included amongst these is the impact of the introduction of passport requirements for American citizens (who, as indicated in the introductory chapter, make up the vast majority of tourists to the Anglophone Caribbean). The Western Hemisphere Travel Initiative (WHTI) is a new U.S. rule which requires American citizens to obtain passports in order to reenter the United States from non-U.S. territories in the Caribbean. Given that only 15 percent of Americans currently hold passports, it seems clear that the WHTI will lead to a decrease in American visitors to the Caribbean, and this has the potential to cost the Caribbean tourism industry more than 188,000 jobs (Yee 2005). Secretary General of the Caribbean Tourism Organisation (CTO) Vincent Vanderpool Wallace has commented that the WHTI is akin to a 'category six hurricane,' while

Caribbean Hotel Association President Peter Odle deemed the move 'catastrophic' (Silvera 2007a).

There are already reports from Jamaica and St. Lucia that the WHTI, which became effective after January 2007, has resulted in double-digit reduction in visitor arrivals out of the United States for the first four months of 2007 (Silvera 2007b). At the time of writing, the CTO is lobbying the U.S. government in order to institute a 'stay of execution' of the WHTI until July 2009, which is the date at which this initiative is expected to be implemented for cruise passengers to the region from the United States (Silvera 2007b). But even if the start date is extended to 2009, this will only be a temporary measure for the Caribbean. Longer term, the Caribbean will need to implement measures to reduce its dependency on the North American tourist market.

Other exogenous factors which have a significant impact on Caribbean tourism include natural disasters, mainly hurricanes; and, as mentioned in the introductory chapter, the most recent such catastrophic event occurred in Grenada in 2004 (Hurricane Ivan), which almost decimated this island's tourism industry. More worryingly, the impact of climate change might result in a more intense hurricane season, sea-level change, and further reef and beach erosion. With regard to climate change, visitors from Europe and the United States are being encouraged to holiday at home in order to reduce their carbon footprints (partly occasioned through extensive air travel), and this has the potential to negatively affect Caribbean tourism. According to David Jessop, Director of the Caribbean Council, the profitability of tourism facilities in the Caribbean is likely to come under increasing pressure from climate-change-related operational costs such as increased insurance and going 'green,' amongst other things (Jessop 2007). The impact of protracted war in Iraq and the threat of terrorist attacks in the key markets of the United States and the United Kingdom also cannot be underestimated and represent challenges to the sustainable competitive advantage of Caribbean tourism.

Evidently, these endogenous and exogenous challenges facing Caribbean tourism are likely to impact on the region's ability to achieve sustainable competitive advantage. This volume cannot lay claim to an exhaustive discussion of these challenges, nor can it claim to have provided definitive solutions. Indeed, any solutions to these challenges must be cognisant of the degree of dynamism within both the internal and external environment, where, to use a rather clichéd expression, the only constant is change. Rather, what we hope this work has done is to propose new perspectives in Caribbean tourism aimed at exposing the multifarious ways of seeing, being in, and experiencing tourism in the region.

But we would be remiss if we did not provide some indication of a future research agenda for new ontologies of Caribbean tourism. While many of these issues have been implicitly interspersed throughout the discussions in this volume, it is useful to comprehensively tie these together in this conclusion. It is recognised that while indigenisation and creolisation are being put forward as integral to the sustainable competitive advantage of the Caribbean tourism product and to new ontologies of tourism in the region, we recognise that care must be taken in our interpretation and operationalisation of these twin processes within the context of tourism. This is largely because of the reality of the ethnic and cultural diversity of many Caribbean societies. Indeed, as Ramcharitar contended in his chapter on tourist nationalism in Trinidad, we must ensure that these twin processes do not themselves serve to privilege some cultures and ethnicities at the expense of others. Further research in this area could involve an exploration of the representation of various cultural and ethnic groupings in Caribbean tourism and the experiences of said groupings within the ubiquitous tourism industry.

The issue of sexuality is another area where discourses of exclusion may be embedded. Indeed, the question may be raised as to the extent to which the issue of sexuality is incorporated in processes of indigenisation and creolisation generally and specifically within a tourism context. Further, it is germane to question if there is recognition within Caribbean societies of expressions of various sexualities and their interaction with and within tourism. This perhaps requires additional probing of the postcolonial critique on homosexuality and tourism instigated by Chambers in this volume.

Another enquiry could focus on how class impacts tourism involvement and experience. In this volume, Brooks focused the discussion of the exclusionary nature of tourist spaces which served to prohibit certain social classes from participating in this vital economic activity. While this discussion was confined to the Caribbean island of Jamaica and to the resort area of Ocho Rios, we can envisage a research agenda which widens this exploration to include other areas within Jamaica and indeed within the wider Caribbean.

The interaction of gender relations and tourism policy and practice is another important area for research, one which has not been directly addressed in this volume. There is need for clarity on how tourism may lead to change in power relationships between men and women, particularly with respect to employment opportunities in the tourism sector. Brooks alluded to this when she noted that unemployed men were more likely to be excluded from the tourist space of the tourism centre of Ocho Rios, Jamaica, than women. The findings of this study revealed that male itinerant traders faced greater prohibition and were more likely to be criminalised than women traders engaged in similar activities. As such, it is important to determine whether there is mainstreaming of gender perspectives in tourism planning and development so that both the perspectives of women and men are included in policy design and implementation in order to benefit both genders and eliminate inequality. Relatedly, a pertinent question is: to what extent have feminist discourses and practices pervaded Caribbean tourism?

The participation of Caribbean diasporic communities (chiefly in America, Canada, and the United Kingdom) in tourism is a further area for exploration. The economic contributions of diasporic communities to the gross national product of many Caribbean countries can scarcely be denied. However, besides remitting money from their adopted countries to their families in the region, there is still a need to determine the extent of the participation of these diasporic communities in tourism activities in the Caribbean and, more importantly, how such participation relates to their sense of identity. Also, to what extent is participation in tourism important to maintaining continued links with their countries of birth?

Finally, as previously indicated, the discussions in this volume have been restricted to the Anglophone or Commonwealth Caribbean and even here to a few of the countries within this context. Consequently, further research should seek to extend this Caribbean tourism critique to other states within the wider Caribbean archipelago and those non-English-speaking territories within the region, which no doubt have different contextual realities.

These are just a few of the areas for further research which we believe are important for future interrogations of Caribbean tourism and which have so far been scarcely explored despite their importance for the discourse and practice of tourism. Ultimately, such investigations are integral to the quest to enhance the competitiveness of Caribbean tourism but, even more crucially, may contribute to wider national and regional strategies to transition from dependency and peripherality to prosperity and self-determination.

It should be noted, however, that the aim of this volume has not been to replace one set of normative views with another of a more home-grown, indigenised variety. It has also not been the intention to negate normative approaches and to impose new prescriptions for Caribbean tourism. Instead, the aim has been to expose the limitations of normative perspectives of Caribbean tourism and in the process engage new ontologies that articulate the developmental realities as well as the aspirations of the peoples of the region.

It is apposite to end this epilogue as it began with another quote from Derek Walcott, who reflects on the attitudes of the West towards the Caribbean:

> . . . how can there be a people there, in the true sense of the word? . . . They know nothing about seasons . . . and are therefore a people incapable of the subtleties of contradiction, of imaginative complexity. So be it. We cannot change contempt . . . (Excerpt from Derek Walcott, Nobel Lecture, December 7, 1992)

This quote seems rather nihilistic with the suggestion that normative discourses about the Caribbean are contemptuous and that there is nothing that can done to foster change. On the contrary, it is hoped that, despite the challenges faced by Caribbean tourism in the twenty-first century, this volume has encouraged a more positive way forward by advocating the creation of more contextual knowledges. However, this is not a call for insularity or a retreat into blinkered introspection. Essentially we are suggesting that 'contemptuous' normative discourses and practices can, and must, indeed be challenged by also 'looking within' as advocated in this volume's emic approach to Caribbean tourism. But this may be just a beginning, or perhaps a moment, or a stage, of what is no doubt a long, precarious, yet hopefully fulfilling journey, in the quest to develop new ontologies of Caribbean tourism. So let the journey continue . . .

REFERENCES

Hartley, J. March 19, 2007. The escape guide: Bite-sized Caribbean. *The Observer,* London. Available online from http://proquest.umi.com/pqdweb?did=10063962 11&sid=1&Fmt=3&clientId=18443&RQT=309&VName=PQD.

Jessop, D. May 13, 2007. Commentary: Tourism limits to growth. *Jamaica Gleaner.*

Silvera, J. January 9, 2007a. 2006 year in review: Great milestones in tourism. *Jamaica Gleaner.*

Silvera, J. June 17, 2007b. Waive visa longer, says Caribbean tourism group. *Jamaica Gleaner.*

Walcott, D. 1997. The Antilles: Fragments of epic memory. In Nobel Lectures, Literature 1991–1995, edited by Sture Allén. Singapore: World Scientific Publishing Co. Available online from www.nobelprize.org.

Yee, A. June 6, 2005. Passport proposal puts dollars 2.6 billion black cloud over the Caribbean. *Financial Times,* London Available online from http://proquest.umi.com/pqdweb?did=1006396211&sid=1&Fmt=3&clientId=18443&RQT=309&VName=PQD.

List of Contributors

Sheere Brooks has worked on medium- to long-term research studies in the areas of poverty, social capital, environmental management, and the politics of local governance both in the Caribbean and the United Kingdom. She is currently completing doctoral research in social policy at the London School of Economics and Political Science and combines her study with working on a number of labour-market and employment studies at the Policy Studies Institute in London. Her research interests are in the areas of Media Communications, Art History, Development Economics and Social Policy, and Planning for developing countries.

Donna Chambers's career in tourism commenced in 1994, when she joined the then Ministry of Industry and Tourism (subsequently the Office of the Prime Minister, Tourism) in Jamaica after completing a master's degree in international relations. As director of tourism administration, her responsibilities included evaluation of hotel developments for government incentives and the drafting of public policies for tourism. She was also initially involved in the development of the first Master Plan for Sustainable Tourism Development for Jamaica. After completing an MSc (distinction) in tourism management from the University of Surrey, she proceeded to complete a Ph.D. tourism with Brunel University, where her research examined the discursive relationship between heritage and the nation and how this relationship was manifested in and through tourism. She is currently Lecturer in Tourism at the University of Surrey, Guildford, UK having previously been employed for five years as a lecturer in tourism and postgraduate programme leader at Napier University, Edinburgh. Her research interests include heritage representation, tourism public policy, postcolonial, and critical theories, and she has presented at several tourism conferences and published in these areas.

Marcella Daye is senior lecturer in tourism and programme manager for the MBA International Tourism at Coventry University in the United Kingdom. She holds a MSc with distinction and a Ph.D. in tourism from the University of Surrey, a postgraduate diploma in mass communications, and an undergraduate degree from the University of the West Indies. Marcella began her career as a radio journalist and then worked as a tourism practitioner for ten years in pubic relations and marketing at the Jamaica Tourist Board. She was also tourism consultant for the National Strategy on Biodiversity in Jamaica, which was commissioned by the UNDP. Marcella has published and presented conference papers in her main research interests of destination image, branding, marketing and consumer behaviour in tourism.

David Dodman is lecturer in the Department of Geography and Geology, University of the West Indies, Mona, where he teaches courses in urban geography, economic geography, and Caribbean geography. He was a Rhodes Scholar at the University of Oxford, where he completed his D.Phil. in 2003. David's research interests are in urban governance, urban livelihoods, Caribbean tourism, and community-based environmental management. His work has been published in journals including *Area, Cities* and *Social and Economic Studies,* and appears in several edited editions.

Wendy C. Grenade holds a Ph.D. and an M.A. in international studies from the University of Miami. She also holds an M.A. in human resource management from the University of Westminster in London. Wendy currently lectures in political science at the University of the West Indies Cave Hill campus. She has several years' experience in public policy and diplomacy, having worked with the Government of Grenada Diplomatic Service (Washington, DC, and London). Her expertise includes multisectoral planning and public-sector reform. Wendy's research interests include regional integration and social development.

Leslie-Ann Jordan holds a Ph.D. in Tourism from the University of Otago. She currently is a lecturer in hospitality management at the University of the West Indies, St. Augustine, Trinidad. She has ten years' experience working in the hospitality and tourism industry in Trinidad and has previously published on institutional arrangements in tourism in small island states of the Caribbean.

Acolla Lewis is a tourism lecturer at the University of the West Indies (UWI), Department of Management Studies, St. Augustine, Trinidad. Her educational background includes an M.Sc. in hospitality and tourism education from the University of Surrey and a Ph.D. in tourism education from Brunel University in the United Kingdom. Her educational pursuits, coupled with her professional experience, have provided her with a sound academic and practical understanding of research, particularly research in the area of tourism development. Acolla has conducted research in tertiary tourism education in the Bahamas, Jamaica, and Barbados. She has also been a principal researcher for the UWI on an UNDP eco-tourism project in Grand Riviere in Trinidad and Tobago.

Jalani Niaah is completing a Ph.D. in cultural studies at the University of the West Indies, Mona. His thesis title is *Rasta Teacher: Leadership, Pedagogy and the New Faculty of Interpretation.* He has published in the *Journal of Cultural Studies* as well as a number of well-based publications. Among the international fora in which he has presented papers are International Cross Roads in Cultural Studies (Birmingham, UK, and Tampere, Finland); Council for the Development of Social Sciences Research in Africa (CODESRIA), General Assembly, Kampala, Uganda; and Conference on Caribbean Culture, UWI, Mona. Niaah's research interests are on popular Jamaican music and culture, Rastafari, and Ethiopianism.

Sonjah Stanley Niaah completed a Ph.D. in cultural studies at the UWI Institute of Caribbean Studies, where she teaches Caribbean cultural theory. She is currently a researcher with the Centre for Caribbean Thought and is also working on two manuscripts: *Kingston's Dancehall: A Story of Space and Celebration* and an edited collection on dance hall (with Bibi Bakare Yusuf, 2005). Sonjah has published on dance hall in *Space and Culture, Discourses in Dance, African Identities,* and *Proudflesh.* Her research interests include urban studies, specifically postcolonial city texts, the performing body, dancehall culture, and Caribbean social psychological knowledge.

Raymond Ramcharitar earned his Ph.D. from the University of the West Indies, St. Augustine. His first book, *Breaking the News: Media and Culture,* was published in 2005 in Trinidad and Tobago. He has been presenting papers on various aspects of Caribbean culture at conferences in the United States and the United Kingdom for the last four years and has worked as a journalist and a lecturer in media and literature in Trinidad for the last 15 years.

Kevon Rhiney is a Ph.D. candidate in the Department of Geography and Geology, University of the West Indies, Mona. His doctoral research investigates the linkages between the local food-supply network and the tourism industry in Jamaica.

Sherma Roberts is an alumnus of the University of the West Indies where she did a first degree in History and a postgraduate diploma in International Relations. She obtained an MSc. Tourism Planning and Development (distinction) from the University of Surrey and her PhD from

Brunel University. Sherma taught in the United Kingdom for one year before returning to the Caribbean, where she is currently.

Douglas Webster is a Jamaican economist of twenty-five years' experience, holding bachelor's and master's degrees in economics from the University of the West Indies, Mona, Jamaica. He is a technology and computer enthusiast with formal training in computer technology. Douglas is also a musicologist with a lifelong passion for popular recorded music, which has in turn inspired continual research and facilitated a broad and deep knowledge of the subject area. He has been collecting shellac and vinyl records for thirty-three years and has a good working familiarity with the vintage-vinyl dealer network.

Index

Printed and bound by CPI Group (UK) Ltd, Croydon, CR0 4YY

01/11/2024

01782630-0012